T0234781

Frontiers in Applied Dynamical Systems: Reviews and Tutorials

Volume 6

More information about this series at http://www.springer.com/series/13763

Frontiers in Applied Dynamical Systems: Reviews and Tutorials

The Frontiers in Applied Dynamical Systems (FIADS) covers emerging topics and significant developments in the field of applied dynamical systems. It is a collection of invited review articles by leading researchers in dynamical systems, their applications and related areas. Contributions in this series should be seen as a portal for a broad audience of researchers in dynamical systems at all levels and can serve as advanced teaching aids for graduate students. Each contribution provides an informal outline of a specific area, an interesting application, a recent technique, or a "how-to" for analytical methods and for computational algorithms, and a list of key references. All articles will be refereed.

Martin Wechselberger

Geometric Singular Perturbation Theory Beyond the Standard Form

 Springer

Martin Wechselberger
School of Mathematics and Statistics
University of Sydney
Sydney, NSW, Australia

ISSN 2364-4532   ISSN 2364-4931   (electronic)
Frontiers in Applied Dynamical Systems: Reviews and Tutorials
ISBN 978-3-030-36398-7   ISBN 978-3-030-36399-4   (eBook)
https://doi.org/10.1007/978-3-030-36399-4

Mathematics Subject Classification: 34E15, 34E17, 34C26, 37Gxx

This Springer imprint is published by the registered company Springer Nature Switzerland AG.
The registered company address is: Gewerbestrasse 11, 6330 Cham, Switzerland

Preface

Many postgraduate students, postdocs and researchers ask me for 'textbook' references on geometric singular perturbation theory and canard theory. Unfortunately, there is 'not much out there'. Notable exceptions are the lecture notes on *Geometric Singular Perturbation Theory* by CKRT Jones [46] and the book on *Multiple Time Scale Dynamics* by C Kuehn [61]. Recently, there have been a couple of reviews, special issues and book chapters on this subject area [15, 22, 25, 113] which reflect the current strong interest in this field of research. Still, there is clearly room for additional expository material.

This manuscript presents a comprehensive review of geometric singular perturbation theory, but with a twist—it focuses on a *coordinate-independent* setup of the theory as presented in Fenichel's original work [30]. The need for such a theory *beyond the standard form* is motivated by looking at biochemical reaction, electronic and mechanical oscillator models that show relaxation-type behaviour. While the corresponding models incorporate slow and fast processes leading to multiple time-scale dynamics, not all of these models take globally the form of a standard slow–fast system. In general, such a standard form where the variables reflect the scale separation can only be achieved locally but *not globally*. Thus from an application point of view, it is desirable to provide tools to analyse singularly perturbed systems in a coordinate-independent manner. Furthermore, this (more) general setting of singularly perturbed problems provides the opportunity for different global (return) mechanisms which allow for different types of relaxation-type behaviour not observed in the standard setting.

The contents cover a general framework for this geometric singular perturbation theory beyond the standard form including canard theory, concrete applications and instructive qualitative models. It contains many illustrations and key points to the existing literature. The target audience are senior undergraduates, graduate students and researchers interested in using the geometric singular perturbation theory toolbox in nonlinear science, either from a theoretical or an application point of view.

Sydney, NSW, Australia   Martin Wechselberger
November 2019

Acknowledgements

This work was supported by the Australian Research Council grant DP180103022. The author would like to thank Peter Szmolyan for helpful feedback on this manuscript. The title of this manuscript *GSPT Beyond The Standard Form* was inspired by the title of lectures given by Peter Szmolyan and the thesis of Ilona Kosiuk.

Contents

Chapter 1
Introduction

Many physical and biological systems consist of processes that evolve on
disparate time- and/or length-scales and the observed dynamics in such systems reflect these multiple-scale features as well. Mathematical models of
such multiple-scale systems are considered singular perturbation problems
with two-scale problems as the most prominent. Singular perturbation theory
studies systems featuring a small perturbation parameter reflecting the scale
separation for which the solutions of the problem at a limiting value of the
perturbation parameter are different in character from the limit of the solutions of the general problem, i.e., the limit is singular. Prandtl's presentation
in 1904 at the third International Congress of Mathematicians in Heidelberg,
Germany, on boundary layers in fluid flows [91] laid the foundation for the
development of singular perturbation theory (with local breakdown of solutions), as well as van der Pol's work in the 1920s on electronic circuits and
relaxation oscillators [106].

This manuscript is concerned with such multiple-scale dynamics and focuses on the geometric approach to singular perturbation theory known as
geometric singular perturbation theory (GSPT) pioneered by Neil Fenichel
in the 1970s [30]. Credit has to go to Andrey Tikhonov as well [104] who
pioneered the field of singular perturbation theory in the Soviet Union; the
interested reader is referred to the survey article by Vasileva [107] about
this development at 'Moscow State University and elsewhere' and references
therein.

In the context of ordinary differential equations (ODEs), singular perturbation problems are usually discussed under the assumption that there exists
a coordinate system such that observed slow and fast dynamics are represented by corresponding slow and fast variables globally, i.e., the system of
ODEs under consideration is given in the *standard (fast) form*

© Springer Nature Switzerland AG 2020

M. Wechselberger, *Geometric Singular Perturbation Theory Beyond
the Standard Form*, Frontiers in Applied Dynamical Systems: Reviews
and Tutorials 4, https://doi.org/10.1007/978-3-030-36399-4_1

$$x' = \varepsilon g(x, y, \varepsilon)$$
$$y' = f(x, y, \varepsilon),$$
(1.1)

where $' = d/dt$ denotes differentiation with respect to the fast time $t \in \mathbb{R}$, or equivalently, the system is given in the *standard (slow) form*

$$\dot{x} = g(x, y, \varepsilon)$$
$$\varepsilon \dot{y} = f(x, y, \varepsilon),$$
(1.2)

where $\dot{} = d/d\tau$ denotes differentiation with respect to the slow time $\tau = \varepsilon t$. In both forms, $x \in \mathbb{R}^k$ represents the slow variables and $y \in \mathbb{R}^{n-k}$ the fast variables with $1 \le k < n$, the vector valued functions f and g are considered to be sufficiently smooth, and $0 < \varepsilon \ll 1$ is the singular perturbation parameter that measures the (global) time-scale splitting between the slow and fast variables. The aforementioned relaxation oscillator model of van der Pol [106] is probably the best-known example of such a singular perturbation problem in standard form; it can be found in many textbooks. GSPT tools to prove the existence of relaxation oscillations are quite intricate and have only been developed recently based on the so-called *blow-up technique* [28, 57]. There exists now a substantial GSPT literature on these problems in standard form, both from the theoretical as well as the application point of view; we refer the reader to the recent book by Kuehn [61] and the many references therein.

It is important to notice that models which incorporate slow and fast processes and show multiple time-scale dynamics do not necessarily have to be of the form (1.1), respectively (1.2). Indeed, as pointed out by Fenichel in his seminal work on GSPT, a global standard *'form is not natural, however, because it depends on the choice of special coordinates'* ([30], page 63). From an application point of view, this raises the important question how to analyse such general multiple time-scale models since standard GSPT cannot be applied directly. Preliminary coordinate transformations to identify and analyse such models as singular perturbation problems in standard form are neither desirable nor, in general, globally possible. Hence, GSPT should be discussed in a more general setting where a given coordinate system does not automatically reflect a global slow-fast variable base. We follow Fenichel's original work [30] and study a system in the general (fast) form

$$z' = H(z, \varepsilon),$$
(1.3)

where $' = d/dt$, or equivalently, a system in the general (slow) form

$$\varepsilon \dot{z} = H(z, \varepsilon),$$
(1.4)

where $\dot{} = d/d\tau$, $\tau = \varepsilon t$, $z \in \mathbb{R}^n$, $n \ge 2$, $H : D \subset \mathbb{R}^n \times \mathbb{R} \to \mathbb{R}^n$ represents a sufficiently smooth vector field in an open subset D, and $0 < \varepsilon \ll 1$ is a singular perturbation parameter, i.e., we assume that system (1.3), respectively, (1.4) shows multiple time-scale dynamics.[1] This manuscript provides a comprehensive review of GSPT in this coordinate-independent setup.

Remark 1.1 *The importance of the standard form* (1.1), *respectively,* (1.2) *is that it represents a* local *canonical form of a general singularly perturbed system* (1.3), *respectively,* (1.4), *i.e., the standard and general singularly perturbed systems are locally topologically equivalent; see Sect. 3.7. Many local results in GSPT have been derived for the standard form of a singularly perturbed system. We will highlight some of these well-known standard local results since they apply (through equivalence) directly to the more general setup discussed in this manuscript.*

The main difference between a standard slow-fast system (1.1) *and a more general slow-fast system* (1.3) *comes through possible different global (return) mechanisms, e.g., the general setting* (1.3) *allows for different classes of relaxation oscillator type behaviour.*

In Chap. 2 we motivate the need for a coordinate-independent approach through specific examples of biochemical reaction networks, electronic circuit and mechanic oscillator models and advection–reaction–diffusion models. In particular, biochemical oscillator models provide a rich source of examples that are not necessarily in standard form, and only recently have such models been studied using GSPT: the mitotic oscillator model by Kosiuk and Szmolyan [52, 54], the autocatalator model [36, 52] and the glycolytic oscillator model [53] by Kosiuk/Gucwa and Szmolyan, the Olson model by Desroches et al. [25] and by Kuehn and Szmolyan [62], and the templator model [13] by Brons, just to name a few.

In Chap. 3, guided by Fenichel's seminal work [30], we review the coordinate-independent GSPT and introduce the slow and fast singular limit problems, the *reduced* and *layer* problems, in this general setting and discuss the (coordinate-independent) normally hyperbolic case known as *Fenichel theory*. We apply this coordinate-independent setup to the normally hyperbolic examples from Chap. 2 and highlight the relationship to the standard case.

In Chap. 4, we then go beyond Fenichel theory and focus on the *loss of normal hyperbolicity* in the context of real eigenvalues. This leads to the definition of *contact points* between the layer and reduced flow and associated *regular jump points* that allow a switch from slow to fast motion in singular perturbation problems. These jump points are an important ingredient for relaxation oscillatory behaviour observed in the motivating examples (Chap. 2), and these examples will be analysed in detail with the coordinate-independent GSPT toolbox in Chap. 5.

The question of the genesis of relaxation oscillations or, more generally, of complex oscillatory pattern generation in singularly perturbed systems is then addressed in Chap. 6 which is closely related to the more degenerate singularities within the set of contact points, known as *pseudo singularities*,

---

[1] This will be made precise in Chap. 3.

and the associated *canard phenomenon* [2, 8, 28, 57]. In particular, we address the concept of excitability in a general singular perturbation problem and the role of canards as excitability threshold manifolds. We highlight the general approach to the canard phenomenon through a modified two-stroke oscillator model that extends the list of motivating examples from Chap. 2.

Finally, in Chap. 7 we provide a brief discussion about additional subject areas related to the coordinate-independent GSPT that have not been covered in this manuscript.

Chapter 2
Motivating Examples

In this chapter we present models that fall under the category of standard singularly perturbation systems (1.1), respectively, (1.2) as well as less known variants of these models that are of the general form (1.3), respectively, (1.4).

2.1 Enzyme Kinetics

Mathematical models of homogeneously mixed reacting systems in chemistry and biochemistry are usually described by systems of ordinary differential equations. Slow and fast reactions in these systems are the norm, and this leads us naturally to singular perturbation problems. A famous and widely studied biochemistry example is given by the *Michaelis–Menten* kinetics scheme (see, e.g., [33, 50, 99] for details),

$$S + E \underset{k_{-1}}{\overset{k_1}{\rightleftharpoons}} C \overset{k_2}{\rightarrow} P + E, \tag{2.1}$$

which models an enzymatic reaction with substrate S, enzyme E, an intermediate complex C and product P. This Michaelis–Menten reaction is assumed to be irreversible, i.e., the product P does not degrade back[1] into the intermediate complex C. Hence, there is no backward arrow included in (2.1) for the reaction $C \overset{k_2}{\rightarrow} P + E$.

Using the *law of mass action* gives the corresponding system of differential equations

---

[1] That is, from a biological point of view the backward reaction is assumed to be significantly slower than all the other reactions involved.

© Springer Nature Switzerland AG 2020
M. Wechselberger, *Geometric Singular Perturbation Theory Beyond the Standard Form*, Frontiers in Applied Dynamical Systems: Reviews and Tutorials 4, https://doi.org/10.1007/978-3-030-36399-4_2

$$\frac{d[S]}{d\tilde{t}} = k_{-1}[C] - k_1[S][E],$$

$$\frac{d[C]}{d\tilde{t}} = k_1[S][E] - (k_{-1} + k_2)[C],$$

$$\frac{d[E]}{d\tilde{t}} = (k_{-1} + k_2)[C] - k_1[S][E], \qquad (2.2)$$

$$\frac{d[P]}{d\tilde{t}} = k_2[C],$$

where $\tilde{t}$ denotes time and $[X]$ denotes the concentration of $X = S, C, E, P$ with initial concentrations

$$[S](0) = S_0, \quad [C](0) = 0, \quad [E](0) = E_0, \quad [P](0) = 0.$$

Notice that $[P]$ can be found by direct integration, and there is a conserved quantity since $d[C]/dt + d[E]/dt = 0$, so that $[C] + [E] = E_0$. Hence it suffices to study the first two equations of system (2.2) with $[E] = E_0 - [C]$.

The first step in our mathematical analysis is to introduce appropriate reference scales for the dependent and independent variables to obtain an associated dimensionless model. Only then we are able to properly identify what are slow and fast processes in the model. As a bonus, the number of effective parameters in the system will be reduced as well. Using such a *dimensional analysis* for system (2.2) gives the corresponding two-dimensional dimensionless system,

$$\frac{ds}{d\tau} = -s + (s + \alpha)c$$

$$\frac{dc}{d\tau} = \beta(s - c(s + \alpha + \gamma)), \qquad (2.3)$$

with dimensionless substrate and complex concentration $s = [S]/S_0 \geq 0$ and $c = [C]/E_0 \geq 0$, dimensionless initial conditions $s(0) = 1$ and $c(0) = 0$, dimensionless time $\tau = E_0 k_1 \tilde{t}$ and three dimensionless parameters[2]

$$\alpha = k_{-1}/(S_0 k_1) \geq 0, \quad \beta = S_0/E_0 \geq 0, \quad \gamma = k_2/(S_0 k_1) \geq 0. \qquad (2.4)$$

Remark 2.1 *The definition of reference scales is, in general, non-unique. This non-uniqueness is often confusing since it might a priori not be clear which is the best choice of reference scales for the problem under study. The basic rule is to create reference scales out of the existing (physical) model*

---

[2] Note the reduction from five physical parameters to three dimensionless parameters.

parameters. Here we use 'obvious' reference scales for the dependent variables [S] and [C]. The reference time scale $(E_0 k_1)^{-1}$ might seem less intuitive, but it reflects a reference 'enzyme reaction' time scale of the model. For an introduction to dimensional analysis, we refer to, e.g., Lin and Segel [71], Chapter 6.

2.1.1 Low Enzyme Concentration

Model Assumption 2.1 *The initial enzyme concentration E_0 is significantly lower than the initial substrate concentration S_0, i.e., $E_0 \ll S_0$. The reaction rates k_2, k_{-1} and $S_0 k_1$ are of the same order.*

Under these model assumptions we have $\alpha = O(1)$, $\beta \gg 1$ and $\gamma = O(1)$ which gives

$$\frac{ds}{d\tau} = -s + (s + \alpha)c$$

$$\varepsilon \frac{dc}{d\tau} = s - (s + \alpha + \gamma)c,$$

(2.5)

where $\varepsilon := \beta^{-1} \ll 1$, i.e., system (2.5) is a singularly perturbed system in the standard (slow) form (1.2) with s as the slow and c as the fast variable. By rescaling the slow time by $\tau = \beta^{-1}t = \varepsilon t$ we obtain the equivalent system

$$\frac{ds}{dt} = \varepsilon(-s + (s + \alpha)c)$$

$$\frac{dc}{dt} = s - (s + \alpha + \gamma)c$$

(2.6)

on the fast time scale t which is a singularly perturbed system in the standard (fast) form (1.1). Figure 2.1 shows the corresponding slow-fast dynamics. After a transient initial time where the complex concentration c quickly builds up while the substrate concentration s stays roughly constant, the observed dynamics become entirely slow for both variables. In the corresponding phase portrait we notice that the slow flow is confined to a curve (a one-dimensional manifold). The slow flow on this manifold approaches the origin which is a stable equilibrium of system (2.6).

The initial transient dynamics can be explained by looking at the singular limit ($\varepsilon \to 0$) of the fast system (2.6) which gives

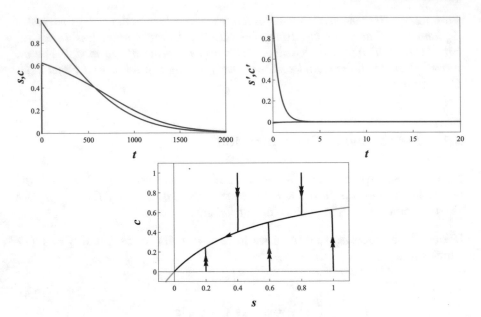

Fig. 2.1 Upper panel: Time traces of the variables (s, c) (left) and zoom of the corresponding time traces of their velocities (right) of classic MM-kinetics model (2.6) with $\alpha = 0.4$, $\gamma = 0.2$, $\varepsilon = 0.01$ and initial condition $(s(0), c(0)) = (1, 0)$. The variable c (red) can be identified as a fast variable relative to the slow variable s (blue). Note the fast initial transient behaviour of c. Lower panel: corresponding phase portrait for different initial conditions (incl. $(s(0), c(0)) = (1, 0)$). After an initial transient fast time, the fast processes do not govern the dynamics of the reaction kinetics which motivates the QSSR

$$\frac{ds}{dt} = 0$$
$$\frac{dc}{dt} = s - (s + \alpha + \gamma)c = f(s, c),$$

$$(2.7)$$

known as the *layer problem* in GSPT. Note that s is 'frozen' which explains why the substrate concentration stays roughly constant for the initial (fast) time of the MM reaction. Only the intermediate complex concentration c shows initial (fast) dynamics. Thus system (2.7) represents a family of one-dimensional dynamical systems for fixed 'parameter' s. The curve or one-dimensional *critical manifold* defined by $f(s, c) = 0$ is the set of equilibria for the layer problem which is (approximately) the curve shown in Fig. 2.1. Note that the partial derivative $\partial f / \partial c = -(s + \alpha + \gamma) < 0$ along this manifold in the physiological relevant domain, the first quadrant, indicating an attractive nature of this 'quasi-steady state' manifold.

This observation is the main motivation behind the widely used *quasi-steady state reduction* (QSSR) technique [99] which assumes that after some transient (fast) initial time, the reaction under study reaches a 'quasi-steady state' and evolves slowly thereafter on this lower dimensional manifold. In our example, one assumes that the intermediate complex c has reached its quasi-steady state (QSS), i.e., we ignore the initial transient fast buildup of the variable c. Formally, we take the limit $\varepsilon \to 0$ in the slow system (2.5) which gives

$$\frac{ds}{d\tau} = -s + (s + \alpha)c$$
$$0 = s - (s + \alpha + \gamma)c = f(s,c),$$
(2.8)

a differential-algebraic problem known as the *reduced problem* in GSPT. Similar to the layer problem, the differential-algebraic problem (2.8) also represents a one-dimensional problem reflecting the singular nature of the original two-dimensional problem, but it is restricted to the critical manifold and, hence, provides complementary information. From a mathematical point of view the question is how do we analyse the reduced problem (2.8), i.e., how do we obtain a vector field in the phase space $\mathbb{R}^2$ restricted to the critical manifold that represents this differential-algebraic problem? Here, we can solve $f(s,c) = 0$ explicitly for c, i.e.,

$$c_0(s) = \frac{s}{s + \alpha + \gamma}.$$
(2.9)

We substitute this expression into (2.8) to obtain

$$\frac{ds}{d\tau} = -s + (s + \alpha)c_0(s) = -\gamma \frac{s}{s + \alpha + \gamma},$$
(2.10)

which describes the evolution of the substrate concentration $s(\tau)$ along the critical manifold. The corresponding complex concentration $c(\tau)$ is then simply found by substituting $s(\tau)$ into (2.9). Equation (2.10) is the QSSR of system (2.5) which is the well-known classic *'Michaelis–Menten' law* that can be found in many textbooks, see, e.g., [50]. The origin is a stable equilibrium of (2.10), i.e., solutions with initial condition $s > 0$ on the critical manifold are attracted towards the origin, which confirms our observation in Fig. 2.1. As we will show in Chap. 3, this leading order approximation of the slow dynamics is rigorously justified by classic GSPT. *Normal hyperbolicity* (defined in Chap. 3) allows here the representation of the critical manifold as a graph over the slow coordinate (chart) s and, hence, the reduction to the one-dimensional dynamics, the Michaelis–Menten law (2.10), globally.

It is important to notice that the slow and fast reactions given through our Model Assumption 2.1 lead directly to a separation into slow and fast variables, i.e., the singularly perturbed system (2.6) is of standard form.

2.1.2 Slow Product Formation

As mentioned previously, a separation into slow and fast variables is by no means a necessary condition for a singularly perturbed system as the following alternative model assumption demonstrates.

Model Assumption 2.2 *The product formation is slow, i.e., the reaction rate k_2 is sufficiently slower than the reaction rates k_{-1} and $S_0 k_1$. The initial enzyme concentration E_0 is comparable to the initial substrate concentration S_0.*

Under these model assumptions we have $\alpha = O(1)$, $\beta = O(1)$ and $\gamma \ll 1$. Hence, we obtain system

$$
\begin{aligned}
\frac{ds}{dt} &= -s + c(s + \alpha) \\
\frac{dc}{dt} &= \beta(s - c(s + \alpha)) - \varepsilon c,
\end{aligned}
\tag{2.11}
$$

with $\varepsilon := \beta\gamma \ll 1$. This system is not in standard form (1.1), but it is clearly a singularly perturbed system as can be seen in Fig. 2.2, i.e., it shows the same features as in Fig. 2.1—fast transient buildup followed by slow motion thereafter. Thus system (2.11) presents a first example of a singularly perturbed system of the general (fast) form (1.3), i.e., neither s or c can be considered a slow variable, both show initial fast motion. This example highlights that model variables in a singularly perturbed system do not have to reflect a slow-fast splitting.

The limit $\varepsilon \to 0$ in (2.11) gives

$$
\begin{aligned}
\frac{ds}{dt} &= -s + c(s + \alpha) \\
\frac{dc}{dt} &= \beta(s - c(s + \alpha)).
\end{aligned}
\tag{2.12}
$$

The singular nature of this layer problem is evident through the existence of a one-dimensional manifold of equilibria implicitly defined by $s - c(s+\alpha) = 0$ and given as a graph

$$
c = c_0(s) = \frac{s}{s + \alpha}.
\tag{2.13}
$$

Evaluating the Jacobian of the layer problem (2.12) along this manifold provides one zero eigenvalue (corresponding to this one-dimensional manifold of equilibria), and one negative eigenvalue, $\lambda_1 = -(\alpha + \beta(s + \alpha)^2)/(s + \alpha) < 0$ in the relevant physiological domain, i.e., the first quadrant, again indicating an attractive nature of this one-dimensional set. Hence, we can assume that this reaction will after some transient (fast) time reach its QSS.

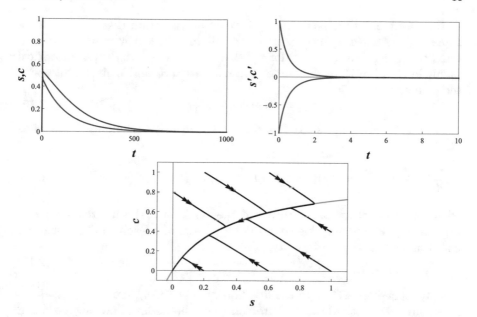

Fig. 2.2 Upper panel: Time traces of the variables (s, c) (left) and zoom of the corresponding time traces of their velocities (right) of general MM-kinetics model (2.11) with $\alpha = 0.4$, $\beta = 1$, $\varepsilon = 0.01$ and initial condition $(s(0), c(0)) = (1, 0)$. Here, both variables show initially a fast transient motion before slowing down, i.e., both have to be considered 'fast' variables. Lower panel: corresponding phase portrait for different initial conditions (incl. $(s(0), c(0)) = (1, 0)$): Note that the fast motion is not aligned with a coordinate axis. Again, after an initial transient time the fast processes do not govern the dynamics of the reaction kinetics which motivates the QSSR

Rescaling system (2.11) to the slow time $\tau = \varepsilon t$ gives

$$\varepsilon \frac{ds}{d\tau} = -s + c(s + \alpha)$$
$$\varepsilon \frac{dc}{d\tau} = \beta(s - c(s + \alpha)) - \varepsilon c,$$

(2.14)

which is a singularly perturbed system in general (slow) form (1.4). If we try to use the 'classic' QSSR technique here, we run into a problem. Taking the singular limit $\varepsilon \to 0$ in (2.14) only implies that we have to restrict the dynamics to the critical manifold (2.13) to define the limit, i.e., the reaction has reached its QSS (2.13). On the other hand, it is not immediately clear how to define a meaningful reduced problem that describes the slow evolution on the critical manifold. So, this provides the first motivation to consider a coordinate-independent approach to GSPT. The main question is how do we obtain the corresponding reduced problem for the limit $\varepsilon \to 0$ in (2.14)?

Here we recall the main idea of singular perturbation theory, i.e., we assume the existence of an (one-dimensional) invariant slow manifold in system (2.11), respectively, (2.14) for $\varepsilon \neq 0$, denoted by S_ε, which has the critical manifold (2.13) as its leading order approximation S_0 and is given as a power series in ε,

$$c(s) = c_0(s) + \varepsilon c_1(s) + \dots, \tag{2.15}$$

where $c_0(s)$ is given by (2.13). Invariance of such a slow manifold S_ε demands that

$$\varepsilon \frac{dc}{d\tau} = \varepsilon c'(s) \frac{ds}{d\tau} = c'(s)(-s + c(s)(s + \alpha)) = \beta(s - c(s)(s + \alpha)) - \varepsilon c(s), \tag{2.16}$$

where $c'(s)$ denotes the derivative of the function $c(s)$ with respect to its argument s. Evaluating this equation by powers of ϵ leads to the first order correction term

$$c_1(s) = -\frac{s}{\alpha + \beta(s + \alpha)^2}. \tag{2.17}$$

Finally, plugging (2.15) into the first equation of (2.14) and taking the limit $\varepsilon \to 0$ gives the desired reduced problem (i.e., the leading slow flow problem)

$$\frac{ds}{d\tau} = \frac{-s(s + \alpha)}{\alpha + \beta(s + \alpha)^2}, \tag{2.18}$$

which describes the evolution of the substrate concentration $s(\tau)$ along the one-dimensional critical manifold (2.13). The corresponding complex concentration $c(\tau)$ is then simply found by substituting $s(\tau)$ into (2.13). Note that the origin is the only equilibrium of (2.18) in the biological relevant domain. It is a stable equilibrium, i.e., solutions with initial condition $s > 0$ on the critical manifold are attracted towards the origin, which confirms our observation in Fig. 2.2.

As will become clear in Chap. 3, this leading order approximation of the slow dynamics can be directly derived from the coordinate-independent GSPT. In particular, we will highlight based on Fenichel's seminal work [30] that identifying an appropriate projection operator provides the geometric tool to calculate the corresponding reduced vector field on the critical manifold S.

Remark 2.2 *A transformation of this general singularly perturbed system (2.11) to standard form is possible here by introducing a new (slow) variable $x = c + \beta s$; note the uniform fast motion shown in the phase portrait, Fig. 2.2. Thus we could derive the above formula via this (global) linear coordinate transformation and then by following the steps of the standard approach, but one does not gain mathematical insight how to derive the reduced flow in the general setup. Hence, we refrain from presenting it.*

2.2 Relaxation Oscillators

Relaxation oscillations are periodic motions that show a mix of slowly and rapidly varying episodes in a cycle—think of the heartbeat, neural firing patterns or blinking lights and electronic beepers as prime examples in our everyday life.

2.2.1 Van der Pol Oscillator

Electrical circuits typically combine capacitative, inductive and resistive elements, and current or voltage sources. Nonlinearities in these circuits may arise through, e.g., transistors, diodes, semi-conductors or operational amplifiers. A famous prototypical oscillating circuit is associated with the Dutch electrical engineer Balthazar van der Pol who studied a circuit with a vacuum tube (within a transistor radio) in the 1920s [105, 106]. Such a circuit for a *van der Pol oscillator* is shown in Fig. 2.3. It is a single loop which has an inductive element, a capacitive element and a tunnel diode in series.[3] Using *Kirchhoff's law* we obtain a second order differential equations for the current $I(t)$ given by

$$\frac{\partial^2 I}{\partial t^2} + \frac{1}{L}f'(I)\frac{\partial I}{\partial t} + \frac{1}{LC}I = 0, \qquad (2.19)$$

where L is the inductance, C the capacitance and $f(I)$ represents the nonlinear current–voltage characteristic of the tunnel diode, i.e., the tunnel diode replaces a passive resistive element with an active resistive element. Such a nonlinear ('active') resistive circuit element allows to pump energy in the

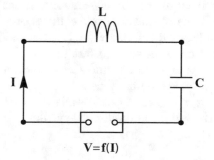

L

I

C

V=f(I)

Fig. 2.3 The van der Pol oscillator circuit with an inductive and capacitive element as well as an active resistive element given by a tunnel diode (or transistor, semi-conductor) with nonlinear current–voltage characteristic $f(I)$. A necessary voltage source (e.g., battery) is omitted in the diagram

---

[3] A necessary voltage source (battery) is omitted in the circuit.

circuit for a specific current regime which provides the means for sustained oscillations in the circuit. For the van der Pol oscillator, $f(I) = \beta I^3 - \alpha I$ is a cubic-shaped nonlinearity which creates negative resistance for small values of the current I but dissipates energy for large values of the current I.

Again, the first step in our mathematical analysis is to introduce appropriate reference scales for the dependent and independent variables to obtain an associated dimensionless model. Using such a *dimensional analysis* for system (2.19) gives the corresponding system

$$\ddot{y} + \mu(y^2 - 1)\dot{y} + y = 0, \tag{2.20}$$

with dimensionless current $y = I/\sqrt{(\alpha/(3\beta))}$, dimensionless time $\tilde{\tau} = \tilde{t}/\sqrt{LC}$ where $\dot{} = d/d\tilde{\tau}$ and a single dimensionless parameter

$$\mu := \frac{\alpha\sqrt{LC}}{L},$$

which measures the ratio of the two characteristic time scales of this circuit given by the characteristic inductive/capacitive time scale, $\sqrt{LC}$, and the characteristic inductive/resistive time scale, L/α.

Model Assumption 2.3 *The characteristic inductive/resistive time scale is significantly faster than the characteristic inductive/capacitive time scale.*

Under this model assumption we have $\mu \gg 1$. We define $\varepsilon := 1/\mu^2$ and obtain

$$\ddot{y} + \frac{1}{\sqrt{\varepsilon}}(y^2 - 1)\dot{y} + y = 0, \tag{2.21}$$

where $\varepsilon \ll 1$. This van der Pol (vdP) model produces sustained *relaxation* oscillations as can be clearly seen in Fig. 2.4, i.e., a mix of slow and fast motions in the time trace of the dimensionless 'current' variable y. In particular, we notice that the relaxation cycle consists of two slow and two fast segments (*'slow-fast-slow-fast'*). Relaxation, i.e., the sudden release of energy that leads to a transition from slow to fast motion, is caused by the small parameter $\varepsilon \ll 1$, i.e., a significant time-scale separation of the interactive elements within the electric circuit model.

If we rescale time by $\tilde{\tau} = \sqrt{\varepsilon}t$ in (2.21), then the vdP relaxation oscillator can be rewritten as the following dynamical system in *Liénard form*:

$$\begin{aligned} x' &= -\varepsilon y \\ y' &= x + y - \frac{y^3}{3}, \end{aligned} \tag{2.22}$$

where $' = d/dt$. This system is a singularly perturbed problem in standard (fast) form (1.1) with x, the dimensionless 'voltage', as slow variable and y, the dimensionless 'current', as fast variable. Looking at the phase portrait

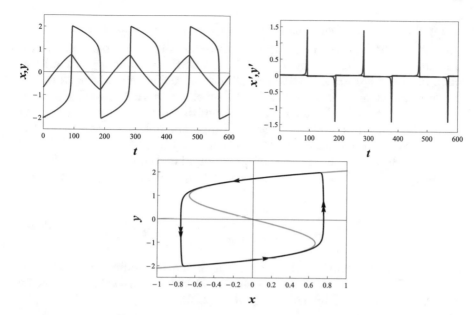

Fig. 2.4 Upper panel: time trace of y (left) and of its velocity (right) of the van der Pol (vdP) relaxation oscillator model (2.21) with $\varepsilon = 0.01$. The blue time trace shows a mix of slow and fast motion (as can be also seen from the red speed trace). Lower panel: the phase portrait of the corresponding system (2.22) which shows that the relaxation cycle forms a hysteresis loop, i.e., the relaxation cycle consists of two slow and two fast segments per cycle (called a four-stroke oscillator by Le Corbeiller [67])

of system (2.22), Fig. 2.4(lower panel), it becomes clear that oscillations are possible due to the non-monotone 'current–voltage' characteristic,

$$x_0(y) = y^3/3 - y \,, \tag{2.23}$$

a necessary condition here for a limit cycle to occur. From a GSPT point of view, the cubic current–voltage characteristic represents the critical manifold of our problem, i.e., it forms the manifold of equilibria of the one-dimensional fast limiting problem, the *layer problem*, obtained by taking the limit $\varepsilon \to 0$ in the fast system (2.22),

$$x' = 0$$
$$y' = x + y - \frac{y^3}{3} = f(x, y) \,. \tag{2.24}$$

As pointed out previously, the layer flow is restricted to the so-called one-dimensional *fast fibres* $\{x = \text{const}\}$. Note that $\partial f/\partial y = (1 - y^2)$ is negative along the outer branches of $x_0(y)$ which identifies them as attractive, while the partial derivative along the middle branch is positive which identifies it as repulsive. At the *fold* points $F^{\pm} = (\mp\frac{2}{3}, \pm 1)$, the partial derivative vanishes which indicates the boundaries of bistability of the two outer attractive branches of $x_0(y)$ denoted by S_a^+ and S_a^-. In GSPT terminology, the critical manifold *loses normal hyperbolicity*[4] at the fold points $F^{\pm}$ where its stability property changes from attracting to repelling or vice versa. Relaxation is induced by the small parameter $\varepsilon \ll 1$ which forces this system to slowly sweep through a hysteresis loop formed (approximately) by segments of the outer attractive branches S_a^+ and S_a^- interspersed by fast switching between these branches at the boundaries of bistability, i.e., the fold points $F^{\pm}$.

Rescaling to slow time $\tau = \varepsilon t$ gives the equivalent system in standard (slow) form,

$$\dot{x} = -y$$
$$\varepsilon \dot{y} = x + y - \frac{y^3}{3},$$

(2.25)

where $\dot{} = d/d\tau$. The slow sweeping through the bistability regime can be identified by looking at the one-dimensional *reduced problem* obtained by taking the limit $\varepsilon \to 0$ in the slow system (2.25),

$$\dot{x} = -y$$
$$0 = x + y - \frac{y^3}{3} = f(x, y).$$

(2.26)

The critical manifold defined by $f(x, y) = 0$ forms the phase space of the reduced problem (2.26) which has a global graph representation over y given in (2.23), i.e., it is a global graph over the fast coordinate (chart) y, not the slow coordinate (chart) x. More importantly, there is no (local) graph representation over the slow coordinate (chart) x in a neighbourhood of the two fold points $F^{\pm}$. Since we would like to understand the slow dynamics near these fold points $F^{\pm}$ where the transition from slow to fast motion happens, we ask again the question how do we obtain a vector field that represents the reduced problem (2.26) on S and, in particular, near the fold points $F^{\pm}$? By definition, such a vector field of the reduced problem has to be tangent to the critical manifold S or, equivalently, orthogonal to the gradient $Df = (\partial f/\partial x, \partial f/\partial y)$ along S, i.e.,

$$\frac{\partial f}{\partial x} \dot{x} + \frac{\partial f}{\partial y} \dot{y} = 0.$$

Since S is represented as a graph over y, the reduced system is given by

$$\dot{y} = -(\frac{\partial f}{\partial y})^{-1} \frac{\partial f}{\partial x} \dot{x} = \frac{y}{1 - y^2},$$

(2.27)

---

[4] Loss of normal hyperbolicity will be discussed in detail in Chap. 4.

which describes the evolution of $y(\tau)$ along the one-dimensional critical manifold S. The corresponding evolution of $x(\tau)$ is then simply found by substituting $y(\tau)$ into (2.23). Firstly, we observe that the reduced flow is towards the fold points $F^{\pm}$. Secondly, the reduced vector field has poles at the fold points, i.e., at $y = \pm 1$, which implies a finite (slow) time blow-up of a solution approaching $F^{\pm}$. The position of the fold points indicates the (approximate) location of the observed switch from slow to fast motion in the relaxation cycle seen in Fig. 2.4(lower panel). Thirdly, the origin is an unstable equilibrium of (2.27) and it is located on the repelling middle branch S_r of the critical manifold which confirms that the reduced flow on S_r is towards each fold point $F^{\pm}$.

Remark 2.3 *We are able to construct a corresponding singular limit ($\varepsilon \to 0$) relaxation oscillator as a concatenation of segments of the two limiting systems, (2.24) and (2.26). GSPT provides then results on the existence of a nearby relaxation oscillator for sufficiently small perturbations $0 < \varepsilon \ll 1$. We will recall the corresponding results in Chap. 4.*

2.2.2 Two-Stroke Oscillator

Let us consider another (non-dimensional) oscillator model,

$$x'' - \left(\frac{\varepsilon}{x' - 1} + x' \right) + x = 0 \,, \tag{2.28}$$

with $\varepsilon \ll 1$ and $' = d/d\tilde{t}$. As in the case of the van der Pol oscillator, we are dealing with a harmonic oscillator with nonlinear damping, but this time it is *velocity-dependent* damping[5] reminiscent of a *Raleigh-type oscillator* model, $x'' - R(x') + x = 0$, with damping characteristic

$$R(x') = \frac{\varepsilon}{x' - 1} + (x' - 1) + 1 \,, \tag{2.29}$$

which is shown in Fig. 2.5 and describes damping relative to the (non-dimensional) speed $v_r = (x' - 1)$. The damping is approximately linear as long as this relative speed v_r stays away from zero, but changes dramatically when this relative speed is approximately zero. The smaller $\epsilon \ll 1$, the more dramatic is the observed change in damping.

Remark 2.4 *We restrict system (2.28) to initial speeds $x_0' < 1$. This guarantees that solutions for this autonomous system exist for all times, i.e., the pole at $v_r = x' - 1 = 0$ does not lead to a finite time blow-up of the corresponding*

---

[5] Which is in contrast to the *displacement-controlled* damping (resistance) in a vdP-type model $y'' + r(y)y' + y = 0$.

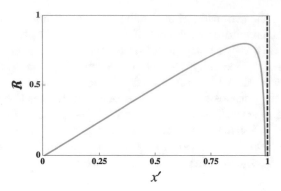

Fig. 2.5 The velocity-dependent characteristic $R(x')$ given by (2.29) of the Raleigh-type oscillator model (2.28); note the sharp transition in the characteristic relative to the speed $v_0 = 1$; compare with a typical friction characteristic of a stick-slip oscillator shown in Fig. 2.7(right)

initial value problem. Figure 2.6 shows an example of an observed two-stroke oscillator time trace. This model 'robustness' will become clear through the following (coordinate-independent) GSPT analysis.

While this might seem to be an atypical nonlinear damping characteristic for a single degree of freedom oscillator model, its dynamics shares strikingly similar features with a simple (dimensionless) *stick-slip* friction oscillator model (see Fig. 2.7),

$$x'' - F(v_r) + x = 0\,, \tag{2.30}$$

a mass subject to a spring force and a velocity-dependent friction force

$$F(v_r) = \begin{cases} x, & v_r = 0, \\ -\mathrm{sgn}(v_r)\mu(v_r), & v_r \neq 0, \end{cases} \tag{2.31}$$

on a constantly moving conveyor belt with speed $v_0 = 1$; see, e.g., [9, 43, 89, 92] for background on friction modelling. The non-negative function $\mu(v_r)$ denotes the *coefficient of friction*. The transition from stick to slip is determined by the *stiction law*, which asserts that the stick phase ($v_r = 0$) is maintained as long as

$$|F(0)| = |x| \leq \mu_s, \tag{2.32}$$

where μ_s denotes the maximal value of static friction capable of preventing the onset of the slipping motion. Note, the friction force $F(v_r)$ is a discontinuous function in a neighbourhood of $v_r = 0$. An example is shown in Fig. 2.7(right).

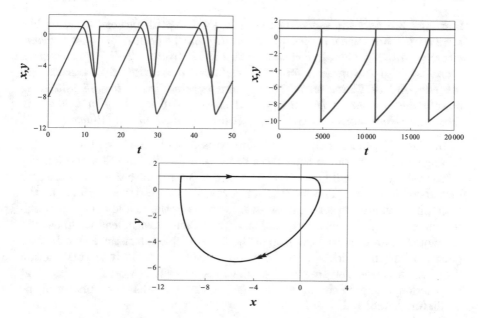

Fig. 2.6 Upper panel: (left) position x (blue) and speed y (red) time trace of the two-stroke oscillator model (2.28) with $\varepsilon = 0.01$; (right) position x (blue) and speed y (red) time trace of the rescaled two-stroke (relaxation) oscillator model (2.36) with $\varepsilon = 0.01$ which shows the slow-fast structure in the time traces. Lower panel: the phase portrait of system (2.36) which shows that the relaxation cycle consists of one slow and one fast segment per cycle (hence, 'two-stroke' oscillator)

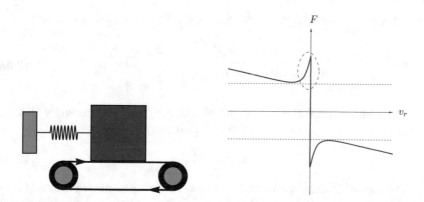

Fig. 2.7 A simple mechanical stick-slip oscillator model (2.30): (left) a mass subject to a spring force and a friction force on a constantly moving conveyor belt; (right) a typical friction force $F(v_r)$. The friction drop near $v_r = 0$ is necessary for stick-slip motion. This is known as the *Stribeck effect*. The green dashed ellipse indicates the domain of interest, where the characteristic R (2.29) smoothly approximates F (for $v_r < 0$); compare with Fig. 2.5

Remark 2.5 *Our toy model* (2.28) *is a local approximation that partially resolves the non-smooth stick-slip oscillator model* (2.30) *for $v_r < 0$ where $F(v_r) = \mu(v_r) = R(v_r)$; compare with Fig. 2.7(right). This approximation for $v_r < 0$ of the friction force is considered a (partial) regularisation of the stick-slip oscillator model* (2.30), *and it is sufficient to understand its autonomous dynamics (for any initial speed $x'(0) < 1$) by means of the (coordinate-independent) GSPT, as will become clear in the following.*

In Fig. 2.6(upper panel, left), we observe a mix of 'slow' and 'fast' motion (changes) in the periodic time trace of system (2.28). The slow motion observed near $x' \approx 1$ could be interpreted as the *'stick'* phase since the conveyor belt speed is $v_0 = 1$. The fast motion away from $x' \approx 1$ corresponds to the *'slip'* phase where the spring takes over harnessing the stored energy through the relaxation mechanism resetting the system and, hence, lead to an oscillator motion. Note the friction drop of the friction characteristic, Fig. 2.5(lower panel), from the 'stick' to the 'slip' phase near $x' \approx 1$. It is this friction drop which causes self-sustained oscillations here. It is worth noticing that our friction/damping characteristic $R(x')$ in (2.29) models both phases of the oscillator, *'stick'* and *'slip'*.

Remark 2.6 *One of the main qualitative differences between this 'stick-slip' type and the vdP oscillator is that the observed limit cycle in this second example consists of only two distinct segments: the* stick *and the* slip *phase. In electronic circuit theory, Le Corbeiller [67] termed such a type of oscillator a* two-stroke oscillator *compared to the classic* four-stroke oscillator *seen in, e.g., the vdP oscillator.*

We recast the two-stroke oscillator model (2.28) as a dynamical system

$$
\begin{aligned}
x' &= y \\
y' &= -x + y + \varepsilon \frac{1}{y-1} \, .
\end{aligned}
\tag{2.33}
$$

Figure 2.6 shows that the behaviour of system (2.33) is markedly distinct whether trajectories are close to the one-dimensional set

$$
y_0(x) = 1 \, ,
\tag{2.34}
$$

where the speed is almost constant. To resolve the dynamics of system (2.33) near this set (which forms a pole of the rational term), we use a dynamical systems tool that is commonly used in GSPT: a state-dependent time-scale transformation,

$$
d\tilde{t} = (1 - y)dt \, ,
\tag{2.35}
$$

which is orientation preserving for $y < 1$ and leads to

$$
\begin{aligned}
x' &= y(1 - y) \\
y' &= (-x + y)(1 - y) - \varepsilon \, ,
\end{aligned}
\tag{2.36}
$$

where with a slight abuse of notation prime refers now to d/dt. System (2.33) and (2.36) are equivalent systems for $y < 1$. System (2.36) is now a singularly perturbed system of the general (fast) form (1.3).

Remark 2.7 *We have topological equivalence between (2.33) and (2.36) for $y < 1$. For $y > 1$, one only has to reverse orientation in (2.36) to obtain the equivalent flow in (2.33). The important insight is that we can study (2.36) in a neighbourhood of the singularity set $\{y = 1\}$.*

The same idea could be used to analyse (2.27): apply the transformation $d\tau = (y^2 - 1)d\tau_1$ to obtain $dy/d\tau_1 = -y$. This system is topologically equivalent to (2.27) for $|y| > 1$ while the flow has to be reversed for $|y| < 1$ to obtain the equivalent flow. State-dependent time-scale transformations such as (2.35) have been the main tool to study special singularities of the reduced problem in GSPT related to loss of normal hyperbolicity; see Chaps. 4 and 6. In the context of models including rational terms with poles, Kosiuk and Szmolyan [53, 54] were the first to apply this rescaling technique.

Taking the singular limit $\varepsilon \to 0$ in (2.36) gives the layer problem

$$\begin{aligned} x' &= y(1 - y) \\ y' &= (-x + y)(1 - y) \,, \end{aligned} \tag{2.37}$$

which has an isolated equilibrium $p = (0,0)$ at the origin which is an unstable focus and a one-dimensional critical manifold S given by (2.34). The Jacobian of system (2.37) evaluated along S has one zero eigenvalue (corresponding to the one-dimensional set S) and a *nontrivial eigenvalue* $\lambda_1 = x - 1$. Hence, S consists of two distinct branches, an attracting branch $S_a = \{(x, y) \in S : x < 1\}$ and a repelling branch $S_r = \{(x, y) \in S : x > 1\}$. Normal hyperbolicity is lost at the *contact point*[6] $F = \{(x, y) \in S : x = 1\}$ where the stability change along S happens.

Looking at the phase portrait of system (2.22), Fig. 2.6(lower panel), we see that the transition from slow to fast motion in the relaxation cycle happens near the contact point F. Relaxation is induced by the small parameter $\varepsilon \ll 1$, but there is no hysteresis loop. Here, it is a fast process itself that resets the oscillator. There is an unstable focus of the layer problem centred at the origin. This isolated singularity $p = (0,0)$ of the layer problem is the distinguishing feature of this 'stick-slip' (or two-stroke) relaxation oscillator compared to the classic van der Pol (four-stroke) relaxation oscillator.

Remark 2.8 *It is this isolated singularity p of the layer problem that prevents a 'global' coordinate transformation of (2.36) to the standard form (1.1). Two-stroke relaxation oscillators form a genuine example of relaxation oscillators without hysteresis.*

---

[6] Contact points will be defined properly in Chap. 4.

Rescaling system (2.36) to the slow time $\tau = \varepsilon t$ gives

$$\varepsilon \dot{x} = y(1 - y)$$
$$\varepsilon \dot{y} = (-x + y)(1 - y) - \varepsilon \,, \tag{2.38}$$

which is a singularly perturbed system in general (slow) form (1.4). The main question is how do we obtain the corresponding reduced problem for the limit $\varepsilon \to 0$ in (2.38)? Clearly, it has to be restricted to the critical manifold S (2.34) which is a graph over x. Again, we show a classical singular perturbation approach first and assume the existence of an (one-dimensional) invariant slow manifold in system (2.36), respectively, (2.38) for $\varepsilon \neq 0$, denoted by S_ε, which has the critical manifold (2.34) as its leading order approximation S_0 and is given as a power series in ε,

$$y(x) = y_0(x) + \varepsilon y_1(x) + \dots, \tag{2.39}$$

where $y_0(x)$ is given by (2.34). Invariance of such a slow manifold S_ε demands that

$$\varepsilon \frac{dy}{d\tau} = \varepsilon y'(x) \frac{dx}{d\tau} = y'(x) y(x)(1 - y(x)) = (-x + y(x))(1 - y(x)) - \varepsilon, \tag{2.40}$$

where $y'(x)$ denotes the derivative of the function $y(x)$ with respect to its argument x. Evaluating this equation by powers of ϵ leads to the first order correction term

$$y_1(x) = \frac{1}{x - 1}.$$

Finally, plugging (2.39) into the first equation of (2.38) and taking the limit $\varepsilon \to 0$ gives the desired reduced problem (i.e., the leading slow flow problem)

$$\frac{dx}{d\tau} = \frac{1}{1 - x}, \tag{2.41}$$

which describes the evolution of $x(\tau)$ along the one-dimensional critical manifold S given by (2.34). We observe that the reduced flow is towards the contact point F (at $x = 1$) and that the reduced vector field has a pole at F which implies a finite (slow) time blow-up of a solution approaching F from S_a (or S_r); see also Remark 2.7 on how to resolve the singularity in (2.41). As mentioned before, the position of F indicates the (approximate) location of the observed switch from slow to fast motion in the relaxation cycle seen in Fig. 2.6(lower panel).

As will become clear in Chap. 4, Fenichel's projection operator method introduced in Chap. 3 can be extended to the case of loss of normal hyperbolicity and thus provide again a coordinate-independent geometric tool to calculate the corresponding reduced vector field near contact points.

Remark 2.9 *In Chap. 5, we construct a corresponding singular limit ($\varepsilon \to$ 0) two-stroke relaxation oscillation as a concatenation of segments of the two limiting systems, (2.37) and (2.41). GSPT provides then results on the existence of a nearby two-stroke relaxation oscillations for sufficiently small perturbations $0 < \varepsilon \ll 1$ and, hence, of two-stroke oscillations in the original model (2.28). A detailed coordinate-independent GSPT analysis of two-stroke oscillator models can be found in [45].*

2.2.3 Three Component Negative Feedback Oscillator

In the design of oscillator models, negative feedback loops provide a possible mechanism for generating robust oscillations. One such *oscillatory motif* is depicted here:

In this motif, $X \to Y$ means 'X activates Y' and $Z \dashv X$ means 'Z inhibits X'. It is considered a *delayed* feedback loop because of three (or more) components connected in a single loop; for a detailed discussion on such biochemical motifs, see, e.g., Novak and Tyson [86]. The following biochemical toy model,

$$\frac{dx}{d\tilde{t}} = \alpha_1 \left(\frac{1}{1+z^2} - x \right)$$

$$\frac{dy}{d\tilde{t}} = \alpha_2 x - \frac{y}{\varepsilon + y} \qquad (2.42)$$

$$\frac{dz}{d\tilde{t}} = \alpha_3 (y - z),$$

adapted from [86], falls under this motif category. It is given in dimensionless form with dimensionless variables $x, y, z \geq 0$ and dimensionless parameters $\alpha_1, \alpha_3 > 0$, $\alpha_2 > 1$ and $\varepsilon \ll 1$. The nonlinear term $1/(1+z^2)$ in the first equation models the inhibition of z on x which is essential for oscillatory behaviour. The degradation of y is modelled by the Michaelis–Menten term $y/(\varepsilon + y)$ which plays a key role in creating relaxation-type behaviour. Figure 2.8(left) shows the oscillatory behaviour found for $\alpha_1 = 0.2$, $\alpha_2 = 2$, $\alpha_3 = 0.2$ and $\varepsilon = 10^{-4}$. The limiting dynamics are very different depending on whether the dynamics are close to $y = O(\varepsilon)$ or not; this is due to the Michaelis–Menten term $y/(\varepsilon + y)$ in the second equation. Similar to the two-stroke oscillator model analysis in Sect. 2.2.2, this singular behaviour is resolved by rescaling time through the state-dependent time transformation $d\tilde{t} = (\varepsilon + y)dt$ which leads to the equivalent system (for $y > 0$)

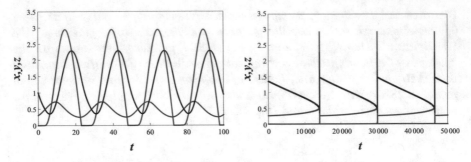

Fig. 2.8 Time traces of three component negative feedback oscillator : x red, y black, z blue. (Left) original biochemical toy model (2.42) with $\alpha_1 = 0.2$, $\alpha_2 = 2$, $\alpha_3 = 0.2$ and $\varepsilon = 10^{-4}$; (Right) the equivalent time rescaled model (2.43) that is in nonstandard form and shows clearly relaxation oscillations

$$\frac{dx}{dt} = x' = \alpha_1\left(\frac{1}{1+z^2} - x\right)(\varepsilon + y)$$
$$\frac{dy}{dt} = y' = \alpha_2 x(\varepsilon + y) - y \tag{2.43}$$
$$\frac{dz}{dt} = z' = \alpha_3(y - z)(\varepsilon + y).$$

It is of the general form (1.3), respectively, (1.4). Figure 2.8(right) shows the time traces of system (2.43) which now clearly indicate a slow-fast time-scale structure in this three component negative feedback biochemical oscillator model.

Remark 2.10 *More complicated relaxation oscillations in biochemical systems (compared to the one shown here) can be found, e.g., in a 3D mitotic oscillator model by Goldbeter [29]. This model has been recast by a state-dependent time transformation (i.e., multiplication of the vector field by the common denominator of all occurring rational functions) as a slow-fast system in general form (1.3) and analysed in detail by Kosiuk and Szmolyan [54] using the GSPT tools presented in this review.*

Taking the singular limit $\varepsilon \to 0$ in (2.43) gives the layer problem

$$\frac{dx}{dt} = x' = \alpha_1\left(\frac{1}{1+z^2} - x\right)y$$
$$\frac{dy}{dt} = y' = (\alpha_2 x - 1)y \tag{2.44}$$
$$\frac{dz}{dt} = z' = \alpha_3(y - z)y,$$

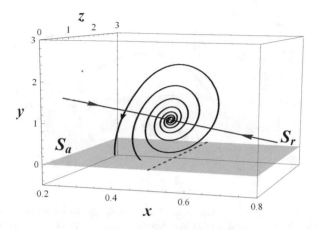

Fig. 2.9 3D phase portrait of the three component negative feedback oscillator with $\alpha_1 = 0.2$, $\alpha_2 = 2$, $\alpha_3 = 0.2$ and $\varepsilon = 0.0001$ showing the saddle-focus isolated singularity P with one-dimensional stable manifold (blue) and two-dimensional unstable manifold

which has an isolated equilibrium $P = (\alpha_2^{-1}, \sqrt{\alpha_2 - 1}, \sqrt{\alpha_2 - 1})^\top$ as well as a two-dimensional critical manifold

$$S = \{(x, y, z) \in \mathbb{R}^3 \; : \; y = 0\}. \tag{2.45}$$

The equilibrium P is of saddle-focus type in the parameter domain of interest with one negative real eigenvalue and a pair of complex conjugate eigenvalues with positive real part[7]; see, e.g., Fig. 2.9.

The Jacobian Dh evaluated along the critical manifold S has two trivial eigenvalues and $\lambda_1 = \alpha_2 x - 1$ as its nontrivial eigenvalue. Thus S loses normal hyperbolicity for $x = \alpha_2^{-1}$, and $S = S_a \cup \tilde{F} \cup S_r$ consists of an attracting branch S_a for $x < \alpha_2^{-1}$ where $\lambda_1 < 0$, a repelling branch for $x > \alpha_2^{-1}$ where $\lambda_1 < 0$ and a one-dimensional submanifold $\tilde{F}$ for $x = \alpha_2^{-1}$ where $\lambda_1 = 0$, i.e., where normal hyperbolicity is lost. It is near this set $\tilde{F}$ where the transition from slow to fast motion in the relaxation cycle happens; see Fig. 2.10.

Rescaling system (2.43) to the slow time $\tau = \varepsilon t$ gives

$$\begin{aligned}
\varepsilon \dot{x} &= \alpha_1 \left(\frac{1}{1 + z^2} - x \right)(\varepsilon + y) \\
\varepsilon \dot{y} &= \alpha_2 x (\varepsilon + y) - y, \\
\varepsilon \dot{z} &= \alpha_3 (y - z)(\varepsilon + y),
\end{aligned} \tag{2.46}$$

which is a singularly perturbed system in general (slow) form (1.4). The main question is how do we obtain the corresponding reduced problem for the limit

---

[7] Calculate the characteristic polynomial for this equilibrium: based on the assumptions on the parameters $(\alpha_1, \alpha_2, \alpha_3)$, *Descartes Sign Rule* shows that there are no real positive roots and the *Routh–Hurwitz Test* shows that not all roots have negative real parts.

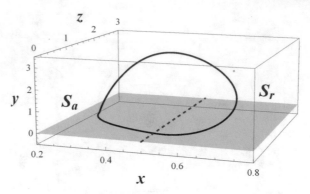

Fig. 2.10 3D phase portrait of the three component negative feedback oscillator with $\alpha_1 = 0.2$, $\alpha_2 = 2$, $\alpha_3 = 0.2$ and $\varepsilon = 0.0001$ showing the two-stroke limit cycle of the three component negative feedback oscillator

$\varepsilon \to 0$ in (2.46)? It has to be restricted to the critical manifold S (2.45) which is a graph over the (x, z) coordinate chart. Again, we show a classical singular perturbation approach first and assume the existence of an (two-dimensional) invariant slow manifold in system (2.46) for $\varepsilon \neq 0$, denoted by S_ε, which has the critical manifold (2.45) as its leading order approximation S_0 and is given as a power series in ε,

$$y(x, z) = y_0(x, z) + \varepsilon y_1(x, z) + \ldots, \qquad (2.47)$$

where $y_0(x, z) = 0$. Invariance of such a slow manifold S_ε demands that

$$\varepsilon \frac{dy}{d\tau} = \varepsilon D_x y \frac{dx}{d\tau} + \varepsilon D_z y \frac{dz}{d\tau}, \qquad (2.48)$$

where $D_x y$ and $D_z y$ denote partial derivatives of the function $y(x, z)$ with respect to its arguments x and z, respectively. Evaluating this equation by powers of ϵ leads to the first order correction term

$$y_1(x, z) = \frac{\alpha_2 x}{1 - \alpha_2 x}.$$

Finally, plugging (2.47) into (2.46) and taking the limit $\varepsilon \to 0$ gives the desired reduced problem (i.e., the leading slow flow problem)

$$
\begin{aligned}
\dot{x} &= \alpha_1 \left(\frac{1}{1 + z^2} - x \right) \left(\frac{1}{1 - \alpha_2 x} \right) \\
\dot{z} &= -\alpha_3 z \left(\frac{1}{1 - \alpha_2 x} \right),
\end{aligned}
\qquad (2.49)
$$

which describes the evolution of $(x(\tau), z(\tau))$ along the two-dimensional critical manifold S given by $y = 0$. This system has an equilibrium at $(x^*, z^*) =$

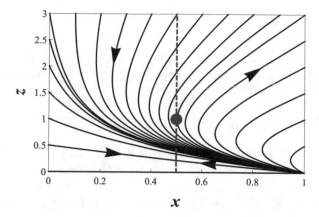

Fig. 2.11 Reduced flow on critical manifold of the three component negative feedback oscillator with $\alpha_1 = 0.2$, $\alpha_2 = 2$, $\alpha_3 = 0.2$

$(1, 0)$ which is an unstable node and lies on the repelling branch S_r—it has no influence on the observed oscillatory behaviour. The reduced flow on the attracting branch S_a is as follows: it is towards the segment of $\tilde{F}$ defined by $0 < z < \sqrt{\alpha_2 - 1}$, while it is away from the opposite half line of $\tilde{F}$ defined by $z > \sqrt{\alpha_2 - 1}$ (this observation is also true for the flow on S_r near $\tilde{F}$); see Fig. 2.11. The reduced vector field has a pole along $\tilde{F}$ which implies a finite (slow) time blow-up of a solution approaching $\tilde{F}$ from S_a (or S_r); see also Remark 2.7 on how to resolve the singularity in (2.49). The position of $\tilde{F}$ indicates the (approximate) location of the observed switch from slow to fast motion in the relaxation cycle seen in Fig. 2.10.

Again, we will show in Chap. 4 that Fenichel's projection operator method introduced in Chap. 3 can be extended to the case of loss of normal hyperbolicity and thus provide a coordinate-independent geometric tool to calculate the corresponding reduced vector field near contact points.

Remark 2.11 *In Chap. 5, based on a contractivity argument of the reduced flow, we construct a corresponding singular limit ($\varepsilon \to 0$) two-stroke relaxation oscillator as a concatenation of segments of the two limiting systems, the layer and the reduced problem. GSPT provides then results on the existence of a nearby two-stroke relaxation oscillator for sufficiently small perturbations $0 < \varepsilon \ll 1$.*

Remark 2.12 *The point $(x, z) = (\alpha_2^{-1}, \sqrt{\alpha_2 - 1}) \in \tilde{F}$ is the location where the reduced flow changes its direction relative to the line $\tilde{F}$. In Sect. 5.2, we will show that this leads to some interesting (local) geometric interpretation related to a* cusp *in singularity theory.*

2.2.4 Autocatalator

One of the simplest models for oscillatory chemical reactions in a closed system can be described by the following autocatalytic feedback motif (see, e.g., [98] for details):

$$P \xrightarrow{k_0} A, \qquad A + 2B \xrightarrow{k_1} 3B, \qquad B \xrightarrow{k_2} C, \qquad A \xrightarrow{k_3} B, \qquad (2.50)$$

with precursor P, intermediates A and B and product C. All reactions are assumed to be irreversible. Hence, there are no backward arrows included in (2.50).

Using the *law of mass action* gives the corresponding system of differential equations

$$\begin{aligned}
\frac{d[P]}{d\tilde{t}} &= -k_0[P], \\
\frac{d[A]}{d\tilde{t}} &= k_0[P] - k_1[A][B]^2 - k_3[A], \\
\frac{d[B]}{d\tilde{t}} &= k_1[A][B]^2 + k_3[A] - k_2[B], \\
\frac{d[C]}{d\tilde{t}} &= k_2[B],
\end{aligned} \qquad (2.51)$$

where $\tilde{t}$ denotes time and $[X]$ denotes the concentration of $X = P, A, B, C$ with initial concentrations

$$[P](0) = p_0, \quad [A](0) = 0, \quad [B](0) = 0, \quad [C](0) = 0.$$

Notice that $[P] = \exp(-k_0\tilde{t})p_0$ and $[C]$ can be found by direct integration. Hence it suffices to study the second and third equations only of system (2.51).

Again, the first step in our mathematical analysis is to introduce appropriate reference scales for the dependent and independent variables to obtain an associated dimensionless model. Using such a *dimensional analysis* for system (2.51) gives the corresponding two-dimensional dimensionless system,

$$\begin{aligned}
\frac{da}{dt} &= -ab^2 - \gamma^2 a + \gamma^2 \mu \exp(-\gamma \delta t) \\
\frac{db}{dt} &= ab^2 - \gamma b + \gamma^2 a,
\end{aligned} \qquad (2.52)$$

with dimensionless concentrations

$$a = [A] \left(\left(\frac{k_2}{k_1} \right) \left(\frac{k_2}{k_3} \right) \right)^{-1/2} \geq 0, \qquad b = [B] \left(\left(\frac{k_2}{k_1} \right) \left(\frac{k_2}{k_3} \right) \right)^{-1/2} \geq 0,$$

dimensionless initial conditions $a(0) = 0$ and $b(0) = 0$, dimensionless time

$$t = \tilde{t}\left(\left(\frac{1}{k_2}\right)\left(\frac{k_3}{k_2}\right)\right)^{-1}$$

and three dimensionless parameters[8]

$$\gamma = \frac{k_3}{k_2} \geq 0, \quad \delta = \frac{k_0}{k_2} \geq 0, \quad \mu = \left(\frac{k_0}{k_2}\right)\left(\frac{k_1}{k_3}\right)^{1/2} \quad p_0 \geq 0. \tag{2.53}$$

Model Assumption 2.4 *The reaction rate k_0 is slow compared to the reaction rate k_3 which in turn is slow compared to the reaction rate k_2. The initial precursor concentration p_0 is large compared to the other chemical concentrations in the model, i.e., $p_0 \gg k_a, k_b$ where $k_a = k_b = ((\frac{k_2}{k_1})(\frac{1}{\gamma}))^{1/2}$ is a typical reference scale of the intermediates A and B introduced above. More precisely, we assume that $p_0 \sim (\frac{\gamma}{\delta})k_a \gg k_a$.*

This model assumption implies that $\delta \ll \gamma \ll 1$ and $\mu = O(1)$. If we restrict our model analysis to time evolutions of τ up to order $O(1/\gamma^2)$ only, then we can 'ignore' the infra-slow dynamics of the large pool precursor and approximate $\exp(-\gamma\delta t) \approx 1$. Applying this approximation[9] we obtain

$$\begin{aligned}\frac{da}{dt} &= -ab^2 - \varepsilon^2 a + \varepsilon^2 \mu \\ \frac{db}{dt} &= ab^2 - \varepsilon b + \varepsilon^2 a,\end{aligned} \tag{2.54}$$

where $\varepsilon := \gamma \ll 1$. This system consists of a mix of $O(1)$, $O(\varepsilon)$ and $O(\varepsilon^2)$ processes.[10] For a range of μ values this system (2.54) shows relaxation oscillatory behaviour as can be seen in Fig. 2.12, and this relaxation cycle also reflects the three time-scales structure as will become apparent in the analysis.

Remark 2.13 *The autocatalator model (2.54) was analysed by Gucwa and Szmolyan [36] using GSPT. We review their work in this manuscript.*

System (2.54) is a singularly perturbed system of the general form (1.3). Taking the singular limit $\varepsilon \to 0$ in (2.54) gives the layer problem

$$\begin{aligned}a' &= -ab^2 \\ b' &= ab^2,\end{aligned} \tag{2.55}$$

which has a critical manifold consisting of two branches, i.e.,

$$S := \{a = 0\} \cup \{b = 0\}. \tag{2.56}$$

---

[8] Note the reduction from five chemical parameters to three dimensionless parameters.

[9] In the chemistry literature, this is known as the *pool chemical approximation*.

[10] We will refer to the corresponding time scales as ultra-fast, fast and slow.

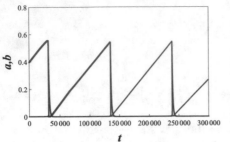 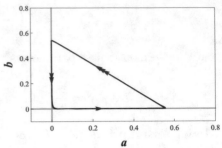

Fig. 2.12 (Left) Time trace of the intermediates a (blue) and b (red) of the autocatalator model (2.54) with $\varepsilon = 0.001$ and $\mu = 6$; (Right) the corresponding phase portrait which shows that the relaxation cycle consists of three segments per cycle reflecting the three time-scales structure of the model (as indicated by the number of arrows: 1 slow, 2 intermediate, 3 fast)

The Jacobian of system (2.55) evaluated along each branch of S has one trivial zero eigenvalue (corresponding to each one-dimensional set of equilibria) and one nontrivial eigenvalue: $\lambda_1 = -b^2$ for the branch $\{a = 0\}$ and $\lambda_1 = 0$ for the branch $\{b = 0\}$. Hence $S_a = \{a = 0 : b > 0\}$ is an attracting branch, while $S_d = \{b = 0\}$ is a degenerate branch of nilpotent equilibria.[11]

The layer flow of system (2.55) is very simple: it evolves along straight one-dimensional fibres $b = -a + b_0$, for any $b_0 > 0$, away from the degenerate branch S_d and towards the attracting branch S_a, as can be seen in Fig. 2.12(right). This explains the ultra-fast upstroke in b shown in Fig. 2.12(left).

Rescaling system (2.54) from the ultra-fast time t to the fast time $\tau = \varepsilon t$ gives

$$\varepsilon \dot{a} = -ab^2 - \varepsilon^2 a + \varepsilon^2 \mu$$
$$\varepsilon \dot{b} = ab^2 - \varepsilon b + \varepsilon^2 a \,, \qquad (2.57)$$

which is a singularly perturbed system in general form (1.4) with $\dot{} = d/d\tau$. The main question is how do we obtain the corresponding reduced problem for the limit $\varepsilon \to 0$ in (2.57)? Clearly, it has to be restricted to one of the branches of critical manifold $S = S_a \cup S_d$.

Let us focus on the attracting branch S_a: again, we show a classical singular perturbation approach first and assume the existence of an (one-dimensional) invariant slow manifold in system (2.54), respectively, (2.57) for $\varepsilon \neq 0$, denoted by $S_{a,\varepsilon}$, which has the critical manifold branch S_a as its leading order approximation $S_{a,0}$ and is given as a power series in ε,

$$a(b) = a_0(b) + \varepsilon a_1(b) + \dots, \qquad (2.58)$$

---

[11] That is, the Jacobian evaluated along the set of equilibria S_d is nilpotent.

where $a_0(b) = 0$. We note that the line $a = 0$ is not only invariant in the singular limit $\varepsilon \to 0$ but also up to $O(\varepsilon)$ perturbations, i.e., there are no $O(\varepsilon)$-terms in the right-hand side of the first equation of (2.57). Thus $a_1(b) = 0$. Plugging this result into the second equation of (2.57) and taking the limit $\varepsilon \to 0$ gives the reduced problem

$$\frac{db}{d\tau} = -b \qquad\qquad (2.59)$$

describing the leading order 'fast' $O(\varepsilon)$ evolution along S_a towards the origin which explains the second phase of the relaxation oscillator shown in Fig. 2.12(left), i.e., the 'fast' downstroke in b (compared to the ultra-fast motion explained previously).

Remark 2.14 *In Chap. 3 we show that the (coordinate-independent) projection operator method provides the means to obtain this leading order approximation of the slow dynamics with GSPT tools.*

It is important to point out that the formal power series expansion of $S_{a,\varepsilon}$ is only valid for $b > 0$ because S_a loses normal hyperbolicity at $b = 0$.

The third and final phase of this autocatalytic relaxation oscillation evolves slowly near $b = 0$, i.e., near the degenerate branch S_d. What is the leading order reduced flow along this branch S_d, i.e., what is the limit $\varepsilon \to 0$ in (2.57)? If we use the same classical approach as above, then we run into several problems[12]:

- the existence of an (one-dimensional) invariant slow manifold in system (2.57) for $\varepsilon \neq 0$ given as a graph $b(a) = b_0(a) + \varepsilon b_1(a) + \dots$ is violated, i.e., there exist two branches/roots of $b_1(a)$;

- the leading order slow flow along S_d is trivial, i.e., $\dfrac{da}{d\tau} = 0$.

The failure of the expansion approach highlighted in the first bullet point is a consequence of the degenerate nature of the branch S_d. The second bullet point should also not come as a surprise since it reflects the inherent three time-scales structure of the processes involved in the autocatalator problem, i.e., the evolution of the relaxation oscillator near $b = 0$ is slow compared to the ultra-fast and fast motions described before. To unravel this third, slow dynamics which are localised near the line $b = 0$, we *zoom in* the neighbourhood of $b = 0$ by rescaling $b = \varepsilon c$ in (2.57), i.e., $c = O(1)$ represents a magnification[13] of $b = O(\varepsilon)$, which gives

$$\begin{aligned}
\dot{a} &= \varepsilon(-ac^2 - a + \mu) = \varepsilon g(a, c) \\
\dot{c} &= ac^2 - c + a = f(a, c),
\end{aligned} \qquad\qquad (2.60)$$

---

[12] This is left as an exercise for the reader.

[13] A proper mathematical justification of this technique known as *blow-up* [16, 27] will be presented in Sect. 5.3.

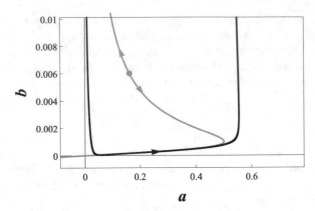

Fig. 2.13 Zoom of the autocatalator model (2.57), Fig. 2.12(right) near $b = 0$, revealing a local standard singular perturbation problem with folded critical manifold

and is a singularly perturbed system in standard (fast) form (1.1), i.e., the new zoomed variable c is fast compared to the slow variable a. This system has a one-dimensional critical manifold $S^z = \{f(a, c) = 0\}$ given as a graph

$$a(c) = \frac{c}{c^2 + 1}. \tag{2.61}$$

Note that $(\partial f / \partial c)|_{a(c)} = (c^2 - 1)/(c^2 + 1)$. Hence, the critical manifold $S^z = S_a^z \cup F^z \cup S_r^z$ consists of an attracting branch S_a^z for $0 < c < 1$ where $(\partial f / \partial c)|_{a(c)}$ is negative and a repelling branch S_r^z for $c > 1$ where $(\partial f / \partial c)|_{a(c)}$ is positive. These two branches meet at a fold point F^z at $c = 1$ where $(\partial f / \partial c)|_{a(c)}$ vanishes and, hence, S^z loses normal hyperbolicity; see Fig. 2.13. Note that the fast motion towards S_a^z at $a = 0$ corresponds to the fast motion along S_b described previously.

Remark 2.15 *As can be seen in Fig. 2.13, S^z is a graph over the c coordinate chart (and not a) which explains the observation made in the first bullet point on page 31. Note that the two branches of the manifold only extend to $a = 1/2$ (compare with the degenerate branch $b = 0$ of S) and that the upper branch asymptotes towards the attracting branch $a = 0$ of S.*

Introducing the slow time scale $\tau_s = \varepsilon\tau$ in (2.60) gives the equivalent system in standard (slow) form,

$$\begin{aligned} \dot{a} &= -ac^2 - a + \mu \\ \varepsilon\dot{c} &= ac^2 - c + a, \end{aligned} \tag{2.62}$$

where, with a slight abuse of notation, $\dot{} = d/d\tau_s$. The observed slow motion along S_a^z in Fig. 2.13 can be identified by taking the limit $\varepsilon \to 0$ in system (2.62),

$$\dot{a} = -ac^2 - a + \mu$$
$$0 = ac^2 - c + a = f(a, c). \tag{2.63}$$

The one-dimensional critical manifold $f(a, c) = 0$ forms the phase space of the reduced problem which has a global graph representation over the coordinate chart c given in (2.61). Recall from the vdP oscillator model analysis, Sect. 2.2.1, a vector field that represents the reduced problem (2.63) on the critical manifold S^z has to be tangent to S^z or, equivalently, orthogonal to the gradient $Df = (\partial f/\partial a, \partial f/\partial c)$ along S^z, i.e.,

$$\frac{\partial f}{\partial a}\dot{a} + \frac{\partial f}{\partial c}\dot{c} = 0.$$

Since the critical manifold S^z is represented as a graph over c (2.61), the reduced system is given in this coordinate chart by

$$\dot{c} = -(\frac{\partial f}{\partial c})^{-1}\frac{\partial f}{\partial a}\dot{a} = \left(\frac{c^2 + 1}{c^2 - 1}\right)(c^3 - \mu(1 + c^2) + c), \tag{2.64}$$

which describes the evolution of $c(\tau_s)$ along the one-dimensional critical manifold S^z. The corresponding evolution of $a(\tau_s)$ is then simply found by substituting $c(\tau_s)$ into (2.61).

Firstly, we observe that the reduced flow has an equilibrium defined by $c^3 - \mu(1 + c^2) + c = 0$ which is given by

$$(a^*, c^*) = (\frac{\mu}{\mu^2 + 1}, \mu). \tag{2.65}$$

This equilibrium[14] is stable for $\mu < 1$ and located on the attracting branch S_a^z, while it is unstable for $\mu > 1$ and located on the repelling S_r^z. To observe relaxation oscillations, this equilibrium has to be unstable and located on the repelling branch S_r^z, i.e., $\mu > 1$ such as shown in Fig. 2.13. In this case, the reduced flow is towards the fold F^z on S_a^z. The reduced vector field has also a pole at the fold F^z given by $c = 1$ which implies a finite (slow) time blow-up of a solution approaching the fold. The position of the fold indicates the (approximate) location of the observed switch from slow to fast motion in the relaxation cycle seen in Fig. 2.13 which explains the termination of the third phase of the relaxation cycle at $a \approx 1/2$ as seen in Fig. 2.12(right).

Remark 2.16 *This simple chemical oscillator model is already remarkably complicated. GSPT and, in particular, the blow-up technique [16, 27] provide the necessary tools to link the dynamics observed for $b = O(\varepsilon)$ with those for $b = O(1)$, which allows to construct a corresponding singular limit $(\varepsilon \to 0)$ relaxation oscillator for $\mu > 1$ as a concatenation of segments of three limiting*

---

[14] Note, it corresponds to an equilibrium in the original system (2.54) given by $(a^*, b^*) = (\frac{\mu}{\mu^2 + 1}, \varepsilon\mu)$.

systems reflecting the three different time scales observed. GSPT provides then results on the existence of a nearby relaxation oscillator for sufficiently small perturbations $0 < \varepsilon \ll 1$. We will recall the corresponding results in Chap. 5. A detailed GSPT analysis of the autocatalator model can be found in Gucwa and Szmolyan [36], which is recommended to the interested reader.

2.3 Advection–Reaction–Diffusion (ARD) Models

Wave fronts are ubiquitous in nature. In the context of population dynamics, such waves may be viewed as representing patterns or structure in migrating populations. Reaction–diffusion equations are used to model population growth dynamics combined with a simple diffusion process to model dispersion and are typically capable of exhibiting travelling wave solutions. In cell migration, advection or convection is another important model mechanism since it may be used to represent tactically driven movement, where cells migrate in a directed manner in response to a concentration gradient [51]. Such migrating cell populations not only form travelling waves but may also develop sharp interfaces in the wave form [44, 65, 88]. Advection, not diffusion, is the main cause for these *shock-type* solutions.

This motivates the study of (dimensionless) nonlinear advection–reaction–diffusion (ARD) models in one spatial dimension, $x \in \mathbb{R}$,

$$u_t + \{g(u)\}_x = f(u) + \varepsilon u_{xx}, \tag{2.66}$$

with $u(x,t) \in \mathbb{R}^n$, $n \geq 1$, and where the diffusion is considered small, i.e., $\varepsilon \ll 1$.

Model Assumption 2.5 *We study system (2.66) with $u \in \mathbb{R}$, $f(u) = u(1 - u)$ and $g(u) = u^2/2$; see Murray [82], pp 458–459.*

Figure 2.14 shows corresponding smooth and shock-type solutions under these model assumptions. In the context of population dynamics, u represents a (non-dimensional) population concentration $u(x,t) \geq 0$, the reaction term $f(u)$ models logistic growth and $Dg(u) = g'(u) = u$ is the concentration dependent advective velocity. The assumption $\varepsilon \ll 1$ emphasises that advection and reaction, not diffusion, are considered the main driving forces in the creation of the observed wave forms including shocks (sharp interfaces). In biochemical reactions, $1/\varepsilon \gg 1$ is reflected by a large Peclét number, a dimensionless quantity that measures the ratio of the advection and diffusion rates, and by a large Damköhler number (of the second kind), another dimensionless quantity that measures the ratio of reaction and diffusion rates.

We are looking for travelling wave profiles $u(x - ct)$ for right moving waves with wave speed $c > 0$ and asymptotic end states $u_\pm$. For (2.66), these asymptotic end states necessarily fulfil $f(u_\pm) = 0$, i.e., the asymptotic end states of a possible travelling wave are given by $u = 0$ and $u = 1$. To identify

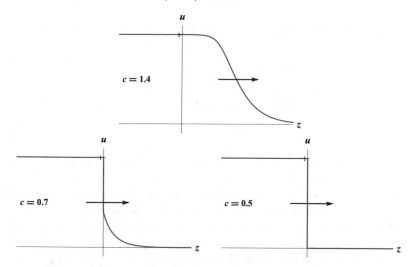

Fig. 2.14 Travelling wave profiles of the ARD model (2.66) in the viscous (diffusion) limit $\varepsilon = 0$: (top) smooth profile for wave speed $c = 1.4$ and (bottom) shock profiles for $c = 0.7$ and $c = 0.5$

profiles $u(x - ct)$ in (2.66), we introduce the travelling wave coordinate $z = x - ct$ which transforms the partial differential equation into a second order ordinary differential equation

$$-c\frac{du}{dz} + \frac{d}{dz}\{g(u)\} = f(u) + \varepsilon\frac{d^2u}{dz^2} \quad \Rightarrow \quad -c\dot{u} + u\dot{u} = u(1 - u) + \varepsilon\ddot{u}, \quad (2.67)$$

where $\dot{} = d/dz$ denotes differentiation with respect to the travelling wave coordinate z. One way to transform this second order differential equation into a dynamical system is to define an auxiliary variable $v = \varepsilon\dot{u}$ which leads to

$$\begin{aligned} \varepsilon\dot{u} &= v \\ \varepsilon\dot{v} &= (u - c)v - \varepsilon u(1 - u). \end{aligned} \quad (2.68)$$

As will become clear shortly, this is a singularly perturbed system of the general (slow) form (1.4). Rescaling the 'slow' independent travelling wave variable $z = \varepsilon y$ in (2.68) gives the equivalent 'fast' system

$$\begin{aligned} u' &= v \\ v' &= (u - c)v - \varepsilon u(1 - u), \end{aligned} \quad (2.69)$$

where $' = d/dy$ denotes differentiation with respect to the 'fast' travelling wave coordinate y. The aim is to find heteroclinic connections between the asymptotic end states $(u_\pm, 0)$ which correspond to travelling waves profiles of the original ARD model (2.66).

Taking the singular limit $\varepsilon \to 0$ in (2.69) gives the layer problem

$$
\begin{aligned}
u' &= v \\
v' &= (u - c)v,
\end{aligned}
\tag{2.70}
$$

which has a one-dimensional critical manifold S given by

$$
v_0(u) = 0 .
\tag{2.71}
$$

Note that the asymptotic end states of a possible travelling wave profile are confined to the critical manifold, i.e., $(u_\pm, 0) \in S$. Calculating the Jacobian of (2.70) and evaluating it along S gives one zero eigenvalue (corresponding to the one-dimensional set S of equilibria) and the nontrivial eigenvalue $\lambda_1 = u - c$. Hence, S consists of two distinct branches, an attracting branch $S_a = \{(u, v) \in S : u < c\}$ and a repelling branch $S_r = \{(u, v) \in S : u > c\}$. Normal hyperbolicity is lost at the *contact point* $F = \{(u, v) \in S : u = c\}$ where the stability change of S happens. Note that solutions of the layer problem (2.70) evolve along parabola

$$
v = u^2/2 - cu + k , \quad k \in \mathbb{R} ,
\tag{2.72}
$$

i.e., this family of parabola parameterised by k represents the fast fibration (foliation) of the layer problem. This indicates that fast connections between the branches S_r and S_a are possible. For the asymptotic end states $u_\pm$ direct fast connections are only possible for the specific wave speed $c = 1/2$ (and $k = 0$) where a connection from $(u_-, v_-) = (1, 0) \in S_r$ to $(u_+, v_+) = (0, 0) \in S_a$ exists.

This specific wave profile of the layer problem for $c = 1/2$ represents a shock in the hyperbolic *balance law* $\{(2.66), \varepsilon = 0\}$—a conservation law with source term. Due to the viscous or dissipative mechanism $(0 < \varepsilon \ll 1)$ in (2.66), these physical shocks are observed as narrow transition regions with steep gradients of the concentration u. Shocks in such hyperbolic balance laws have to fulfil the *Rankine–Hugoniot (RH) condition*. For (2.66) this is given by

$$
c[u] = [g(u)],
\tag{2.73}
$$

where $[u] = u_+ - u_-$ and $[g(u)] = g(u_+) - g(u_-) = (u_+^2 - u_-^2)/2$ denote the discontinuous jumps in these quantities across the shock. The RH condition (2.73) can be rewritten as

$$
cu_- - g(u_-) = cu_+ - g(u_+) = k ,
\tag{2.74}
$$

where $k \in \mathbb{R}$ is a parameter. This RH condition is fulfilled for the asymptotic end states $(u_-, v_-) = (1, 0)$ and $(u_+, v_+) = (0, 0)$ for the specific wave speed $c = 1/2$ only. This confirms that shocks have to occur between points on the critical manifold S which are connected by fast fibres as described by (2.72). The unidirectional connection from $u_- \in S_r$ to $u_+ \in S_a$ encoded by the stability properties of the branches of the critical manifold S is equivalent to a *Lax entropy* condition of the shock in the PDE context.

Remark 2.17 *For background on hyperbolic PDE theory we refer the reader to, e.g., Courant and Hilbert [20] and Lax [66].*

To find possible other connections between the asymptotic end states $u_\pm$, we have to understand the corresponding slow dynamics. We could either look for slow connections between the asymptotic end states or connections that are a concatenation of slow and fast segments. The main question is how do we obtain the corresponding reduced problem for the limit $\varepsilon \to 0$ in (2.68)? Clearly, it has to be restricted to the critical manifold S. Again, we show a classical singular perturbation approach first and assume the existence of an (one-dimensional) invariant slow manifold in system (2.68) for $\varepsilon \neq 0$, denoted by S_ε, which has the critical manifold (2.71) as its leading order approximation S_0 and is given as a power series in ε,

$$v(u) = v_0(u) + \varepsilon v_1(u) + \dots, \tag{2.75}$$

where $v_0(u)$ is given by (2.71). Invariance of such a slow manifold S_ε demands that

$$\varepsilon \frac{dv}{d\tau} = \varepsilon v'(u) \frac{du}{d\tau} = v'(u)v(u) = (u - c)v(u) - \varepsilon u(1 - u), \tag{2.76}$$

where $v'(u)$ denotes the derivative of the function $v(u)$ with respect to its argument u. Evaluating this equation by powers of ϵ leads to the first order correction term

$$v_1(u) = \frac{u(1 - u)}{u - c}.$$

Finally, plugging (2.75) into the first equation of (2.68) and taking the limit $\varepsilon \to 0$ gives the desired reduced problem (i.e., the leading slow flow problem)

$$\frac{du}{d\tau} = \frac{u(1 - u)}{u - c}, \tag{2.77}$$

which describes the evolution of $u(\tau)$ along the one-dimensional critical manifold S given by (2.71). The two asymptotic end states of a possible travelling wave, $u_- = 1$ and $u_+ = 0$, are necessarily equilibria of the reduced problem. The equilibrium $u_+ = 0$ is stable independent of the wave speed $c > 0$, while the stability property of the other equilibrium $u_- = 1$ depends on $c > 0$—it is stable for $0 < c < 1$ and unstable for $c > 1$. Note further that for $c > 1$, both equilibria are on the attracting branch S_a. Hence, for $c > 1$ we identify smooth travelling wave profiles connecting from $u_- = 1$ to $u_+ = 0$, i.e., right moving travelling waves that invade the unoccupied state $u_+ = 0$.

In the case $0 < c < 1$, both equilibria are stable for the reduced problem but $u_- = 1$ is on the repelling branch S_r of the critical manifold. Hence, for fixed $0 < c < 1$, the only possibility to connect the two end states is by jumping from the asymptotic end state $u_- = 1$ to the attracting branch connecting to $u_l = 2c - 1 < 1$. From this intermediate point u_l the reduced flow connects to the other asymptotic end state $u_+ = 0$ (this is not necessary for $c = 1/2$ since the fast fibre connects directly to $u_l = u_+ = 0$ as mentioned above). If we restrict to $u \geq 0$, then we can only find wave profiles with shocks for $1/2 \leq c < 1$. Example of wave profiles with shocks is shown in (u, v) phase space in Fig. 2.15.

Remark 2.18 *In Chap. 4 we provide the geometric tool to calculate the corresponding reduced vector field near contact points.*

Remark 2.19 *For $c > 1$, we identify smooth wave profiles entirely on S_a described by the reduced flow only. GSPT provides then results on the existence of a nearby travelling wave for sufficiently small perturbations $0 < \varepsilon \ll 1$; see Chap. 3.*

For $0 < c < 1$, we are able to construct a corresponding singular limit ($\varepsilon \to 0$) travelling wave profile with a shock as a concatenation of segments of the two limiting systems, (2.70) and (2.77); see Fig. 2.15. Again, GSPT provides then results on the existence of a nearby travelling wave for sufficiently small perturbations $0 < \varepsilon \ll 1$; see Chaps. 3 and 4.

Remark 2.20 *We do not discuss the (PDE) stability question of such travelling wave profiles. We conjecture that none of the travelling waves is (PDE) stable, but that we find physical travelling waves with wave speed $c \geq 1/2$ in (2.66) under Model Assumption 2.5; these is based on similar observations presented in [40].*

For an excellent introduction to the subject area on the stability of travelling waves, we refer to, e.g., Sandstede [96].

Remark 2.21 *We would like to point out that the travelling wave problem (2.67) of an ARD system (2.66) can always be transformed to a singular perturbation system in standard form as follows: rewrite (2.67) as*

$$\frac{d}{dz}\{\varepsilon\frac{du}{dz} + cu - g(u)\} = -f(u) \tag{2.78}$$

and define

$$v := \varepsilon\frac{du}{dz} + cu - g(u). \tag{2.79}$$

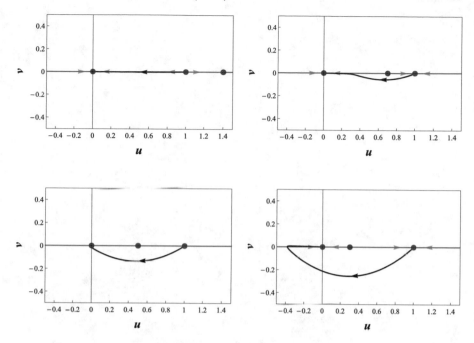

Fig. 2.15 Travelling wave profiles (heteroclinic connections) including shocks in (u, v) phase plane. Blue dots represent the end states of the wave, while the red dot indicates the contact point F of the critical manifold S; (upper left) wave speed $c = 1.4$, (upper right) wave speed $c = 0.7$; (lower left) wave speed $c = 0.5$, (lower right) wave speed $c = 0.3$

This leads to a singularly perturbed system in Liénard *form,*

$$\dot{v} = -f(u) \tag{2.80}$$
$$\varepsilon\dot{u} = v - cu + g(u),$$

where $\dot{} = d/dz$ *which is in standard (slow) form* (1.2). *We refer to [112] where this approach was introduced and [38, 39] for corresponding model studies on tumour invasion and wound healing angiogenesis.*

Chapter 3
A Coordinate-Independent Setup for GSPT

This chapter is devoted to present a geometric approach to singular perturbation theory for ordinary differential equations. The material is based on Fenichel's seminal work on *geometric singular perturbation theory* with a particular emphasis on his coordinate-independent approach (see [30], Sections 5–9).

Remark 3.1 *This exposition on Fenichel's coordinate-independent GSPT summarises also more recent work by Goeke, Noethen and Walcher, see, e.g., [34, 85]. Their coordinate-independent work on 'Tikhonov Fenichel' reductions was motivated by the quasi-steady state phenomenon which occurs frequently for differential equations that model biochemical reactions such as presented in Sect. 2.1.*

Consider the general system (1.3), i.e.,

$$z' = H(z, \varepsilon), \tag{3.1}$$

where $z \in \mathbb{R}^n$, $n \geq 2$, $H : U \times I \subset \mathbb{R}^n \times \mathbb{R} \to \mathbb{R}^n$, is a sufficiently smooth vector field, $' = d/dt$ denotes differentiation with respect to $t \in \mathbb{R}$ and $\varepsilon \in I = (0, \varepsilon_0)$ with $\varepsilon_0 \ll 1$ is a small perturbation parameter[1]. We note that the vector field $H(z, \varepsilon)$ could have additional parameter dependence, but we only make this explicit where necessary.

Assumption 3.1 *The vector field $H(z, \varepsilon)$ of (3.1) has the following power series expansion in ε,*

---

[1] More precisely, Fenichel [30] assumes that H is a C^r function, $r \geq 2$, defined on an open subset $U \times I \subset \mathbb{R}^n \times \mathbb{R}$, where $I = (-\varepsilon_0, \varepsilon_0)$ with $\varepsilon_0 \ll 1$, i.e., negative $\varepsilon < 0$ are included as well. In applications, we usually restrict to positive ε and assume that H and its derivatives up to order r have limits as $\varepsilon \to 0+$.

© Springer Nature Switzerland AG 2020
M. Wechselberger, *Geometric Singular Perturbation Theory Beyond the Standard Form*, Frontiers in Applied Dynamical Systems: Reviews and Tutorials 4, https://doi.org/10.1007/978-3-030-36399-4_3

$$H(z, \varepsilon) = h(z) + \varepsilon G(z, \varepsilon), \qquad (3.2)$$

where $h(z)$ is the leading order term and $\varepsilon G(z, \varepsilon)$ is the remainder of this series.

Due to the small parameter $\varepsilon \ll 1$, system (3.1) defines a perturbation problem which could be either regular or singular depending on the structure of the set of singularities of h denoted by S_0, i.e.,

$$S_0 := \{ z \in \mathbb{R}^n \: : \: H(z, 0) = h(z) = \mathbb{O}_n \}, \qquad (3.3)$$

where $\mathbb{O}_n$ denotes an n-dimensional (column) vector with zero entries.

Definition 3.1 *System* (3.1) *is a* regular *perturbation problem if the set S_0* (3.3) *is either empty or it consists entirely of isolated singularities.*

Definition 3.2 *System* (3.1) *is a* singular *perturbation problem if there exists a subset $S \subseteq S_0$ of* (3.3) *which forms a k-dimensional differentiable manifold, $1 \leq k < n$. The set S is called the* critical manifold *of system* (3.1).

Remark 3.2 *A k-dimensional differentiable manifold S is locally diffeomorphic to $\mathbb{R}^k$, i.e., there exists a collection of coordinate charts (U_i, ϕ_i) known as an* atlas[2], *where the U_i are open sets covering S and each $\phi_i : U_i \to \mathbb{R}^k$ is a diffeomorphism from U_i to an open subset $\phi(U_i) \subset \mathbb{R}^k$. All transition maps $\phi_{ji} = \phi_j \circ \phi_i^{-1}$ restricted to $\phi_i(U_i \cap U_j)$ are diffeomorphisms as well.*

The graph of a smooth function $F : U \subseteq \mathbb{R}^k \to \mathbb{R}^{n-k}$, which is a subset of $\mathbb{R}^k \times \mathbb{R}^{n-k}$ defined by $\Gamma(F) = \{ (x, y) \in \mathbb{R}^k \times \mathbb{R}^{n-k} \: : \: x \in U, \, y = F(x) \}$, is an example of a smooth manifold. More precisely, the graph $\Gamma(F)$ is a smooth embedded submanifold of the manifold $\mathbb{R}^k \times \mathbb{R}^{n-k} = \mathbb{R}^n$.

A smooth level set of a smooth function $\Phi : U \subseteq \mathbb{R}^n \to \mathbb{R}$ is defined as the set $M = \Phi^{-1}(c)$, $c \in \mathbb{R}$. If the gradient $D\Phi(a)$ is nonzero for all $a \in \Phi^{-1}(c)$, then the set M forms a smooth $(n-1)$-dimensional embedded submanifold of $\mathbb{R}^n$.

For an excellent introduction to smooth manifolds we refer the reader to, e.g., the textbook by Lee [68].

Assumption 3.2 *System* (3.1) *is a singular perturbation problem with a single open subset $S \subseteq S_0$ of* (3.3) *which forms a k-dimensional connected differentiable manifold of singularities, $1 \leq k < n$. This set $S \subseteq S_0$ is the k-dimensional zero level set of the smooth function $h : U \subset \mathbb{R}^n \to \mathbb{R}^n$ with constant rank $(n-k)$ for all $z \in S$, i.e., S is an embedded submanifold.*

Remark 3.3 *In general, the set of singularities S_0 could be a union of disjoint manifolds or a union of manifolds intersecting along lower dimensional submanifolds. There could be also isolated singularities within the set S_0. In fact, this is very common in applications.*

---

[2] Since different atlases can cover the same manifold, an equivalence class of compatible atlases has to be defined; see, e.g., [68, 101].

*The autocatalator model presented in Sect. 2.2.4 'violates' Assumption 3.2,
i.e., the set $S = S_a \cup S_d$ (2.56) with $S_a \cap S_d \neq \emptyset$, and S_d is not a regular
level set either, but it can be analysed using GSPT and the blow-up method
[36]; see Sect. 5.3. Again, this degeneracy is very common in applications and
reflects the multiple time-scale structure (more than two).*

3.1 The Layer Problem

We study system (3.1) as a singular perturbation problem evolving on the
fast time scale t called the *fast* system.

Definition 3.3 *The leading order fast problem obtained in the singular limit
$\varepsilon \to 0$ of system (3.1), i.e.,*

$$z' = H(z,0) = h(z) \,, \tag{3.4}$$

is called the layer problem.

Under Assumption 3.2, we focus on the case of a connected critical manifold S which is the k-dimensional zero level set of the smooth function
$h : U \subset \mathbb{R}^n \to \mathbb{R}^n$ with constant rank $(n-k)$ for all $z \in S$. It follows
that the Jacobian Dh evaluated along S has at least k zero eigenvalues corresponding to the k-dimensional tangent space $T_z S$ of S at each point $z \in S$.
We call these k eigenvalues *trivial*. The remaining $(n-k)$ eigenvalues denoted
by λ_i, $i = 1, \ldots, n-k$, are called *nontrivial*. Clearly, the geometric multiplicity of the zero eigenvalue along S is fixed to k because the dimension of
the manifold S is fixed by its definition, i.e., the Jacobian $Dh|_S$ must have
full rank $(n-k)$, while the algebraic multiplicity of the zero eigenvalue may
change, i.e., it could be greater than or equal k depending on the (continu-
ous/smooth) variation of the nontrivial eigenvalues along S.

Definition 3.4 *Let $S_n \subseteq S$ denote the subset where all nontrivial eigenvalues
of Dh evaluated along S_n are nonzero and let $S_h \subseteq S_n$ denote the subset where
all nontrivial eigenvalues of Dh evaluated along S_h have nonzero real part.*

Remark 3.4 *The set $S_h \subseteq S$ is k-dimensional. If the subset $S \backslash S_h$ exists,
then it is lower dimensional. In general, it forms a (union of) submanifold(s)
of codimension-one.*

Since $h(z)$ vanishes identically along S, the tangent space $T_z S$ is in the kernel
of $Dh(z)$ and this k-dimensional subspace is invariant under the map $Dh(z)$.
This invariant subspace $T_z S$ induces a linear map on the $(n-k)$-dimensional
quotient space $T_z \mathbb{R}^n / T_z S$, i.e., $T_z \mathbb{R}^n / T_z S$ is also invariant under $Dh(z)$ and
its corresponding eigenvalues are the nontrivial eigenvalues.

This linear map on the quotient space $T_z \mathbb{R}^n / T_z S$ is invertible if all $\lambda_i \neq$
0, $i = 1, \ldots, n-k$, i.e., for $z \in S_n$. For each $z \in S_n$ the tangent space

$T_z S$ coincides with the kernel of $Dh(z)$ since the geometric and algebraic multiplicity of the zero eigenvalue is the same—it is k. In this case, the tangent space $T_z S$ has for all $z \in S_n$ a unique complementary invariant subspace denoted by N_z, i.e., we have the splitting

$$T_z \mathbb{R}^n = T_z S \oplus N_z, \quad \forall z \in S_n. \tag{3.5}$$

We denote the $(n - k)$-dimensional subspace N_z as a (linear) *fast fibre* with *base point* $z \in S_n$ which is a concrete realisation of the quotient space $T_z \mathbb{R}^n / T_z S$.

Definition 3.5 *The set $TS_n = \cup_{z \in S_n} T_z S_n$ forms the* tangent bundle *of S_n and the set $N = \cup_{z \in S_n} N_z$ is the corresponding (linear)* fast fibre bundle[3].

Remark 3.5 *With the splitting (3.5), one can associate a (unique) projection operator*

$$\Pi^{S_n} : T\mathbb{R}^n|_{S_n} = TS_n \oplus N \to TS_n,$$

which defines the map from $T\mathbb{R}^n$ onto the base space TS_n along N which is part of the definition of a (smooth) fibre bundle structure due to the product space splitting given in (3.5).

Let $f(z) = (f_1(z), \ldots, f_{n-k}(z))^\top$ denote a column vector of sufficiently smooth functions $f_i : \mathbb{R}^n \to \mathbb{R}$, $i = 1, \ldots, n - k$, such that the k-dimensional connected differentiable manifold of equilibria $S \subseteq S_0$ is defined as the *regular level set*

$$S = \{z \in \mathbb{R}^n : f(z) = \mathbb{O}_{n-k}\}, \tag{3.6}$$

i.e., the $(n - k) \times n$ Jacobian $Df(z)$ has full (row) rank for all $z \in S$.

Assumption 3.3 *The function $h(z)$ can be factored as follows:*

$$h(z) = N(z)f(z) \tag{3.7}$$

with $n \times (n - k)$ matrix $N(z)$ formed by column vectors $N^i(z) = (N_1^i(z), \ldots, N_n^i(z))^\top$ with sufficiently smooth functions $N^i : \mathbb{R}^n \to \mathbb{R}$, $i = 1, \ldots, n-k$. The matrix $N(z)$ has full (column) rank for all $z \in S$. Singularities of $N(z)f(z)$ for $z \notin S$, if they exist, are isolated[4].

Remark 3.6 *Not all embedded submanifolds can be expressed as level sets of smooth submersions, i.e., not every singularly perturbed system (3.1) with a smooth k-dimensional connected critical manifold S can be factorised globally in the form (3.7). Even in the case of a polynomial (or rational) vector*

---

[3] In differential geometry it is usually referred to as the *transversal bundle*. For a general introduction to fibre bundles; see, e.g., [68, 101].

[4] This assumption on the singularities of $N(z)f(z)$ for $z \notin S$ could be weakened to being simply disjoint from the set S; see also Remark 3.3.

field $h(z)$, one cannot expect that such a factorisation (3.7) holds globally. However, every embedded submanifold is at least locally of this form [68], and there is an algorithmic/constructive approach that utilises this splitting locally as will become clear in the following; see, e.g., [34, 85].

Lemma 3.1 *Given the k-dimensional zero level set $S_n \subseteq S$ of the sufficiently smooth function $h(z)$ with constant rank $(n-k)$, i.e., all nontrivial eigenvalues of $Dh|_S$ are nonzero. Then the factorisation (3.7) must hold locally, and the column vectors of $N(z)$ form a basis of the range of the Jacobian $Dh(z)$, while (the transpose of) the row vectors of $Df(z)$ form a basis of the orthogonal complement of the kernel of $Dh(z)$ at each base point $z \in S_n$.*

Proof The Jacobian $Dh(z)$ evaluated along $z \in S_n$ has rank $(n - k)$. Thus there exist $(n - k)$ linearly independent column vectors in $Dh(z)$ for $z \in S_n$. Let $\{N^1(z), \ldots, N^{n-k}(z)\}$ denote a basis of this $(n - k)$-dimensional column space of $Dh(z)$, i.e., of the linear fibre $\mathcal{N}_z$, and let $N(z) = [N^1(z) \cdots N^{n-k}(z)]$ be the corresponding $n \times (n-k)$ matrix. Hence, every column vector of $Dh(z)$ can be represented as a linear combination of this basis. In matrix form, this implies that $Dh(z) = N(z)F(z)$ for all $z \in S_n$, where $F(z)$ is a $(n - k) \times n$ matrix with full row rank $(n - k)$.

Let $\{W^1(z), \ldots, W^k(z)\}$ denote a basis of the kernel of $Dh(z)$ evaluated at $z \in S_n$, i.e., a basis of the tangent space $T_z S$, and let $W(z) = [W^1(z) \cdots W^k(z)]$ be the corresponding $n \times k$ matrix. We have

$$Dh(z)W(z) = N(z)F(z)W(z) = \mathbb{O}_{n,k} \implies F(z)W(z) = \mathbb{O}_{n-k,k} \quad \forall z \in S_n,$$

where $\mathbb{O}_{n,k}$ denotes an $n \times k$ matrix with zero entries, since the column vectors of $N(z)$ form a basis of the column space. Hence, (the transpose of) the $n - k$ row vectors of $F(z)$ form a basis of the orthogonal complement of the k-dimensional kernel of $Dh(z)$ at each $z \in S_n$, i.e., of $T_z S^\perp$.

Recall, a k-dimensional differentiable manifold is (locally) defined by (3.6) if the $(n-k) \times n$ Jacobian $Df(z)$ has full rank for all $z \in S_n$. In this case, (the transpose of) the corresponding $n - k$ independent row vectors of $Df(z)$ form a basis of $T_z S^\perp$. Thus we set $F(z) := Df(z)$ and obtain $Dh(z) = N(z)Df(z)$ along $z \in S_n$. Hence, the factorisation (3.7) follows locally for all $z \in S_n$. $\square$

Since $h(z)$ vanishes identically along the critical manifold S, the layer problem (3.4) defines the corresponding nontrivial $(n - k)$-dimensional (fast) flow which describes locally the (fast) motion near the k-dimensional critical manifold S. Denote by $\mathcal{B} \subset \mathbb{R}^n$ a sufficiently small tubular neighbourhood of the critical manifold S such that the layer problem (3.4) has no additional singularities in $\mathcal{B}$ besides S, i.e., $N(z)$ has full (column) rank in $\mathcal{B}$. Let $L(z) = (L_1(z), \ldots, L_k(z))^\top$ be a column vector of sufficiently smooth functions $L_i : \mathbb{R}^n \to \mathbb{R}$, $i = 1, \ldots, k$, and let

$$\mathcal{L}^c = \{z \in \mathbb{R}^n : L(z) = c \in \mathcal{U} \subset \mathbb{R}^k\} \tag{3.8}$$

define $(n - k)$-dimensional regular level sets parameterised by $c \in \mathcal{U} \subset \mathbb{R}^k$.

Lemma 3.2 *Given local $(n - k)$-dimensional regular level sets $\mathcal{L}^c$ (3.8) in a tubular neighbourhood $\mathcal{B} \subset \mathbb{R}^n$ of the critical manifold S, i.e., $\mathrm{rk}\, DL|_{\mathcal{B}} = k$. If $DL(z)N(z)f(z) = \mathbb{O}_k$ holds for all $z \in \mathcal{B}$, then these local $(n-k)$-dimensional level sets $\mathcal{L}^c$ form a local invariant nonlinear 'foliation' $\mathcal{L} = \cup_{c \in \mathcal{U}} \mathcal{L}^c$ of S for the layer problem*

$$z' = h(z) = N(z)f(z) \, . \tag{3.9}$$

Each nonlinear fibre $\mathcal{L}^c$ that has a corresponding base point $z \in S$ is tangent to its linear fibre $\mathcal{N}_z$, i.e., $DL(z)N(z) = \mathbb{O}_{k,n-k}$ holds for all $z \in S$.

Proof A set is locally invariant under the flow of the layer problem (3.9) in a neighbourhood $\mathcal{B} \subset \mathbb{R}^n$ of S if for any initial condition $z_0 \in \mathcal{L}^c \cap \mathcal{B}$ the corresponding trajectory $z(t) \in \mathcal{L}^c \cap \mathcal{B}$ for all $t \in \mathbb{R}$, i.e., $L(z(t)) = c$, unless $z(t)$ leaves the set $\mathcal{L}^c$ through the boundary $\partial \mathcal{L}^c = \mathcal{L}^c \cap \partial \mathcal{B}$ in forward or backward time. Invariance under the layer flow (3.9) implies the defining condition $DL(z)z' = DL(z)N(z)f(z) = \mathbb{O}_k$ for $\mathcal{L}^c$. Note that $N(z)f(z) \neq \mathbb{O}_n$ for all $z \in \mathcal{B}\backslash S$.

For $z \in S$, Lemma 3.1 shows that the set $\{N^1(z), \ldots, N^{n-k}(z)\}$ forms a basis of the $(n - k)$-dimensional nontrivial linear flow of $Dh(z)$, i.e., it forms the span of each fast linear fibre $\mathcal{N}_z$. Thus $DL(z)N(z) = \mathbb{O}_{k,n-k}$ at each $z \in S$ is a necessary condition for the corresponding nonlinear fibre $\mathcal{L}^c$ to be tangent to $\mathcal{N}_z$, i.e., it provides a smooth continuation of the set $\mathcal{L}^c$ from $\mathcal{B}\backslash S$ to its corresponding base point $z \in S$ (where it exists). $\qquad\square$

Remark 3.7 *Lemma 3.2 holds for all $z \in S$, i.e., even in those cases where some elements of $\mathcal{L}$ do not necessarily have a corresponding local base point $z \in S$ and, hence, the notion of an invariant 'foliation' is not justified. Nonetheless, the set $\mathcal{L}$ still forms a local invariant 'foliation' of the layer problem. Furthermore, Assumption 3.3 guarantees that the span of the set $\{N^1(z), \ldots, N^{n-k}(z)\}$ forms an $(n - k)$-dimensional subspace for all $z \in S$. Thus we can extend the nonlinear 'foliation' $\mathcal{L}$ by continuity passed lower dimensional subsets $S\backslash S_n$ of the critical manifold where the corresponding nonlinear 'fibres' have a tangency with $S\backslash S_n$; see Chap. 4.*

Remark 3.8 *Only in the case $k = n - 1$, the condition $DL(z)N(z)f(z) = \mathbb{O}_{n-1}$ for all $z \in \mathcal{B}$ is equivalent to $DL(z)N(z) = \mathbb{O}_{n-1}$ for all $z \in \mathcal{B}$. Hence for $k = n - 1$, $N(z)$ represents the vector field of the layer problem by equivalence through the time transformation $dt_1 = f(z)dt$.*

Lemma 3.3 *Given the layer problem (3.9) of a singularly perturbed system in general form*

$$z' = H(z, \varepsilon) = N(z)f(z) + \varepsilon G(z, \varepsilon) \tag{3.10}$$

with k-dimensional (connected) critical manifold $S \subseteq S_0$. The nontrivial eigenvalues of the corresponding Jacobian evaluated along the critical manifold S (3.6) are the eigenvalues of the $(n - k) \times (n - k)$ matrix $DfN|_S$.

Proof We straighten locally the critical manifold S: define local coordinates $z = (x, y)^{\top}$, $x \in \mathbb{R}^k$ and $y \in \mathbb{R}^{(n-k)}$, such that $D_y f(x, y)$ is a regular $(n - k) \times (n - k)$ square matrix, and the $n \times (n - k)$ matrix $N = (N^x, N^y)^{\top}$ with $k \times (n - k)$ matrix N^x and $(n - k) \times (n - k)$ square matrix N^y. It follows from the *implicit function theorem* that the critical manifold S, i.e., the regular level set $f(x, y) = 0$, is locally given as a graph $y = Y(x)$ over the x-coordinate chart. Hence we can define a local coordinate transformation,

$$v = f(x, y), \tag{3.11}$$

with $y = K(x, v)$ as its inverse which transforms (3.9) locally to

$$\begin{pmatrix} x' \\ v' \end{pmatrix} = \begin{pmatrix} N^x(x, K(x, v)) \\ DfN(x, K(x, v)) \end{pmatrix} v. \tag{3.12}$$

The corresponding Jacobian $\tilde{J}$ evaluated along S is given by

$$\tilde{J}|_S = \begin{pmatrix} \mathbb{O}_{k,k} & N^x(x, K(x, v)) \\ \mathbb{O}_{n-k,k} & DfN(x, K(x, v)) \end{pmatrix}. \tag{3.13}$$

Hence, the nontrivial eigenvalues are the eigenvalues of the $(n - k) \times (n - k)$ matrix DfN evaluated along S (which is independent of the choice of local coordinates). $\qquad\square$

The properties of the nontrivial eigenvalues of $Dh|_S$ provide a classification scheme for singular perturbation problems.

Definition 3.6 *A k-dimensional (sub)manifold $S_h \subseteq S$ is called* normally hyperbolic[5] *if the $(n - k) \times (n - k)$ matrix $DfN|_{S_h}$ is hyperbolic. Such a manifold S_h is called* attracting *if all nontrivial eigenvalues have negative real parts,* repelling, *if all nontrivial eigenvalues have positive real parts or* saddle-type *otherwise.*

Remark 3.9 *GSPT that is concerned with a normally hyperbolic critical manifold S_h is referred to as* Fenichel theory *[30, 46].*

In the case of a normally hyperbolic manifold S_h, we split the nonlinear foliation $\mathcal{L}$ into sub-foliations that reflect fast motion towards or away from S_h.

Definition 3.7 *The local stable and unstable manifolds of a normally hyperbolic critical manifold S_h, denoted by $W^s_{loc}(S_h)$ and $W^u_{loc}(S_h)$, respectively, are the (disjoint) unions*

---

[5] For $z \in S_h$, the transverse bundle $\mathcal{N}$ can be transformed locally into a *normal* bundle which explains the nomenclature; see Lemma 3.6.

$$W_{loc}^s(S_h) = \bigcup_{z \in S_h} W_{loc}^s(z), \quad W_{loc}^u(S_h) = \bigcup_{z \in S_h} W_{loc}^u(z), \tag{3.14}$$

where $W_{loc}^s(z)$ and $W_{loc}^u(z)$ are local stable and unstable invariant manifolds of the base point $z \in S_h$ of the layer problem (3.4). They form a family of locally invariant nonlinear fast fibres for $W_{loc}^s(S_h)$ and $W_{loc}^u(S_h)$, respectively, that are tangent to the stable linear sub-fibres N_z^s and unstable linear sub-fibres N_z^u, respectively, of N_z.

Remark 3.10 *The local manifold $W_{loc}^s(S_h)$ is $(k + n_s)$-dimensional and $W_{loc}^u(S_h)$ is $(k + n_u)$-dimensional, where $n_s + n_u = n - k$. These local invariant manifolds $W_{loc}^s(S_h)$ and $W_{loc}^u(S_h)$ can be extended by the layer flow (3.4) to global invariant manifolds $W^s(S_h)$ and $W^u(S_h)$.*

Remark 3.11 *Under Assumption 3.3, singularities $p \in S_0/S$ of $N(z)$, if they exist, are isolated (and bounded away from S). If such an isolated singularity $p \in S_0/S$ is hyperbolic, i.e., the corresponding Jacobian has all its eigenvalues bounded away from the imaginary axis, then this hyperbolic singularity will persist in system (3.10) for $0 < \varepsilon \ll 1$, i.e., we are dealing locally with a regular perturbation problem; compare with Definition 3.1.*

3.2 The Reduced Problem

In system (3.1), respectively, (3.10), if we change from the fast time scale t to the slow time scale $\tau := \varepsilon t$, then we obtain the equivalent (slow) system in general form, i.e.,

$$\dot{z} = H(z, \varepsilon) = \frac{1}{\varepsilon} N(z) f(z) + G(z, \varepsilon), \tag{3.15}$$

evolving on the slow time scale τ, i.e., $\dot{} = d/d\tau$. In system (3.15), the singular limit $\varepsilon \to 0$ is only well defined if

- the phase space $\mathbb{R}^n$ is restricted to the k-dimensional critical manifold S where $f(z)$ vanishes and
- the leading order perturbation vector field $G(z, 0) \in T_z\mathbb{R}^n$ is restricted to its component in the corresponding tangent space $T_z S$ for each base point $z \in S_n$.

To obtain such a reduced vector field that describes the (reduced) flow on S to leading order explicitly, we use the fact that there exists a unique projection operator $\Pi^S : T\mathbb{R}^n \to TS$ onto the tangent bundle TS defined for all base points $z \in S_n$ associated with the (linear) fast fibre bundle N defined by the splitting (3.5), i.e., $T\mathbb{R}^n = TS \oplus N$ along S_n; see Remark 3.5.

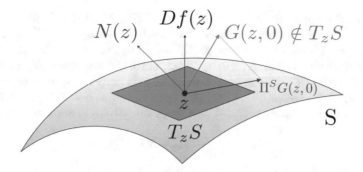

Fig. 3.1 The construction of the reduced vector field of a general singularly perturbed system (3.15) through the projection operator Π^S

Definition 3.8 *The singular limit system of (3.15) is given by*

$$\dot{z} = \Pi^S \left. \frac{\partial}{\partial \varepsilon} H(z, \varepsilon) \right|_{\varepsilon=0} = \Pi^S G(z, 0) \,, \tag{3.16}$$

where the vector field $\Pi^S G(z, 0)$ is defined on the tangent bundle TS for each base point $z \in S_n$. System (3.16) is called the reduced problem *and the vector field $\Pi^S G(z, 0)$ induces the k-dimensional reduced flow on S_n; see Fig. 3.1.*

The following recalls the notion of an *oblique projection* due to a given splitting of a vector space:

Definition 3.9 *Given the splitting $\mathbb{R}^n = \mathcal{V} \oplus \mathcal{W}$, where $\dim \mathcal{V} = n - k$ and $\dim \mathcal{W} = k$, $1 \leq k < n$. Let V, respectively, U be the $n \times (n - k)$ matrices whose column vectors span $\mathcal{V}$, respectively, $\mathcal{W}^\perp$. The* oblique projection *of a vector $\xi \in \mathbb{R}^n$ onto the subspace $\mathcal{V}$ parallel to $\mathcal{W}$ is defined by*

$$\Pi^{\mathcal{V}} \xi = V(U^\top V)^{-1} U^\top \xi \,. \tag{3.17}$$

The complementary *oblique projection of a vector $\xi \in \mathbb{R}^n$ onto the subspace $\mathcal{W}$ parallel to $\mathcal{V}$ is defined by*

$$\Pi^{\mathcal{W}} \xi = (\mathbb{I}_n - V(U^\top V)^{-1} U^\top) \xi \,. \tag{3.18}$$

Lemma 3.4 *Given a singularly perturbed system in general form (3.15). Then the corresponding reduced problem (3.16) is given by*

$$\dot{z} = \Pi^S G(z, 0) = \left(\mathbb{I}_n - N(Df\, N)^{-1} Df \right) G(z, 0) \tag{3.19}$$

with $z \in S_n$ and $\mathbb{I}_n$ denotes the n-dimensional identity matrix.

Proof In the splitting (3.5), i.e., $T_z\mathbb{R}^n = T_z S \oplus \mathcal{N}_z$, the subspace $\mathcal{V} = \mathcal{N}_z$ is spanned by the column vectors of N and the subspace $\mathcal{W}^\perp = T_z S^\perp$ is spanned by the column vectors of $Df^\top$. Hence we can define the oblique projection of the vector $G(z,0) \in T_z\mathbb{R}^n$ onto the subspace $\mathcal{N}_z$ parallel to $T_z S$ by formula (3.17). The complement of this projection in $T_z\mathbb{R}^n$, i.e., the projection of the vector $G(z,0) \in T_z\mathbb{R}^n$ onto the subspace $T_z S$ parallel to $\mathcal{N}_z$, is then defined by formula (3.18) and gives our desired result with projector

$$\Pi^S := \mathbb{I}_n - N(Df\, N)^{-1} Df\,, \tag{3.20}$$

i.e., $(\Pi^S)^2 = \Pi^S$ and $\mathrm{rk}(\Pi^S) = k$ for all $z \in S_n$. $\qquad\square$

Remark 3.12 *In his seminal work [30], Fenichel presents an equivalent formula of the projection operator in Lemma 5.4. Similarly, Goeke and Walcher present such an equivalent formula in [33], Theorem 8.2.*

Assumption 3.4 *Singularities of the reduced problem* (3.19), *defined by*

$$\Pi^S G(z,0) = \left(\mathbb{I}_n - N(Df\, N)^{-1} Df\right) G(z,0) = \mathbb{O}_n\,, \tag{3.21}$$

are isolated for $z \in S_n$.

There are two possibilities to obtain singularities in (3.21).

- Firstly, the perturbation vector $G(z,0) \in T_z\mathbb{R}^n$ itself has singularities for $z \in S_n$. Assumption 3.4 then implies that these singularities of $G(z,0)$ are isolated.
- Secondly, the projection vector $\Pi^S G(z,0) \in T_z S$ has singularities for $z \in S_n$, but not $G(z,0) \in T_z\mathbb{R}^n$ itself. The projection $\Pi^S G(z,0)$ (3.21) defines at most k independent equations for the existence of singularities on the k-dimensional critical manifold S. Assumption 3.4 picks the number of these independent equations to be exactly k.

Remark 3.13 *There is also the possibility of an m-dimensional subset of singularities of the reduced vector field $\Pi^S G \in TS$, $1 \le m < k$, which would indicate a multiple time-scale structure with more than two time scales in the underlying singular perturbation problem* (3.15).

Note that in the autocatalator model the projector Π^s itself is not well defined along the (sub-)set S_d because S_d is not a regular level set due to a vanishing Jacobian; see Sects. 2.2.4 and 5.3.

Remark 3.14 *The reduced problem* (3.16) *holds for all $z \in S_n$, i.e., also for $z \in S_n \backslash S_h$ where normal hyperbolicity is lost. In Chap. 4, we discuss the reduced problem* (3.16) *in a neighbourhood of the complementary set $S \backslash S_n$ where normal hyperbolicity is also lost.*

3.3 A Slow-Fast Example in General Form

Given

$$x' = -2(x^2 + y - 1)x + 2\varepsilon$$
$$y' = -(x^2 + y - 1)y - \varepsilon x + \varepsilon^2 \tag{3.22}$$

which is a singularly perturbed system in the general form (3.10) with

$$z = \begin{pmatrix} x \\ y \end{pmatrix} \in \mathbb{R}^2, \ N(x,y) = -\begin{pmatrix} 2x \\ y \end{pmatrix}, \ f(x,y) = x^2 + y - 1, \tag{3.23}$$

and

$$G(x, y, \varepsilon) = \begin{pmatrix} 2 \\ -x + \varepsilon \end{pmatrix}. \tag{3.24}$$

Layer Problem:
The set of singularities of the layer problem

$$\begin{pmatrix} x' \\ y' \end{pmatrix} = -\begin{pmatrix} 2x \\ y \end{pmatrix} (x^2 + y - 1) \tag{3.25}$$

is given by

$$S_0 = \{(0,0)^\top\} \cup \{x^2 + y - 1 = 0\}.$$

The isolated singularity at the origin defined by $N(x,y) = 0$ is an unstable node. The subset $S \subset S_0$ forms the one-dimensional critical manifold

$$S = \{(x,y) \in \mathbb{R}^2 \ : \ x^2 + y - 1 = 0\} \tag{3.26}$$

of this singular perturbation problem. Here, the critical manifold S is a regular level set and it can be represented as a graph $y_0(x) = 1 - x^2$ over the single coordinate chart $x \in \mathbb{R}$. The Jacobian Dh evaluated along $(x, y) \in S$ is given by

$$Dh|_S = \begin{pmatrix} -4x^2 & -2x \\ 2x(x^2 - 1) & x^2 - 1 \end{pmatrix}$$

which has a trivial zero eigenvalue and $\lambda_1 = Df N|_S = -(4x^2 + y)|_S = -(3x^2 + 1) < 0$, $\forall x \in \mathbb{R}$, as its nontrivial eigenvalue. Hence by Definition 3.6, $S = S_h = S_a$ is an attracting normally hyperbolic manifold.

Here, the corresponding (global) stable nonlinear fast fibres of the layer problem can be calculated explicitly and are given as a family of curves

$$W^s(S_h) = \bigcup_{z \in S_h} W^s(z) = \bigcup_{c \in \mathbb{R}} \{x = cy^2\} \cup \{y = 0\}.$$

All these nonlinear fibres terminate at the isolated singularity $(0, 0)^\top \in N$ (which is $\notin S$); see Fig. 3.2.

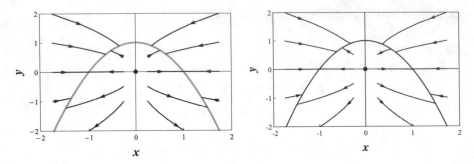

Fig. 3.2 Phase portrait of system (3.23): (left) layer problem and (right) full problem. The critical manifold S, the parabola is normally hyperbolic and the reduced flow on S is uniformly increasing

Reduced Problem:
Here, DfN is a scalar and its value evaluated along S corresponds to the single nontrivial eigenvalue λ_1 of the layer problem. We obtain the following reduced problem (3.19),

$$\begin{pmatrix} \dot{x} \\ \dot{y} \end{pmatrix} = \left(\begin{pmatrix} 1 & 0 \\ 0 & 1 \end{pmatrix} + \frac{1}{4x^2 + y} \begin{pmatrix} -4x^2 & -2x \\ -2xy & -y \end{pmatrix} \right) G(x, y, 0)$$

which has to be evaluated along the critical manifold S. Since S is a graph over the x-coordinate chart, $y_0(x) = 1 - x^2$, we study the reduced flow in this single coordinate chart which leads to

$$\dot{x} = 2 - \frac{6x^2}{1 + 3x^2} = \frac{2}{1 + 3x^2} > 0$$

through our specific choice of $G(x, y, 0)$ given in (3.24). Hence, the reduced flow on S is simply uniform slow motion and there exist no isolated singularities of the reduced problem.

If we replace (3.24) by a different perturbation vector

$$G(x, y, \varepsilon) = \begin{pmatrix} 2 \\ 3x + \varepsilon \end{pmatrix} \tag{3.27}$$

then the corresponding reduced problem (3.19) becomes

$$\dot{x} = 2 - \frac{14x^2}{1 + 3x^2} = \frac{2 - 8x^2}{1 + 3x^2}$$

which has two isolated singularities at $x_\pm = \pm \frac{1}{2}$ with x_- being repulsive and x_+ being attractive.

Remark 3.15 *Clearly, the existence and number of isolated singularities cannot be deduced from the structure of the perturbation vector G alone, e.g.,* *both choices of perturbation vectors $G(x, y, 0)$ in (3.24) and (3.27) have no singularities themselves, but only the projection $\Pi^S G(x, y, 0)$ of (3.27) gives singularities in the reduced problem; see discussion below Assumption 3.4.*

Under Assumption 3.4, the perturbation vector $G = (G_1(x, y, 0), G_2(x, y, 0))^\top$ has the property that its projection onto TS does not vanish identically along S, i.e., $(1-x^2)G_1(x, y, 0) - 2xG_2(x, y, 0)$ is not identical to zero, which implies the existence of a nontrivial reduced flow.

3.4 Enzyme Kinetics with Slow Production Rate Revisited

Recall system (2.3) which is a dimensionless description of the Michaelis–Menten reaction kinetics (2.1) under the Model Assumption 2.2 that the product formation is significantly slower than the other reaction rates. This model (2.3) is a singularly perturbed system in general form with

$$z = \begin{pmatrix} s \\ c \end{pmatrix} \in \mathbb{R}^2, \ N(s, c) = \begin{pmatrix} -1 \\ \beta \end{pmatrix}, \ f(s, c) = -s + (s + \alpha)c, \qquad (3.28)$$

and

$$G(s, c, \varepsilon) = \begin{pmatrix} 0 \\ -c \end{pmatrix}. \qquad (3.29)$$

Layer Problem:

The set of singularities of the layer problem

$$\begin{pmatrix} s' \\ c' \end{pmatrix} = \begin{pmatrix} -1 \\ \beta \end{pmatrix} (-s + (s + \alpha)c) \qquad (3.30)$$

is given by

$$S = \{-s + (s + \alpha)c = 0\} \qquad (3.31)$$

which forms the one-dimensional critical manifold of this problem (there are no isolated singularities). Here, the critical manifold S is a regular level set and it can be represented as a graph $c_0(s) = s/(s + \alpha)$ over the single coordinate chart $s \in \mathbb{R}_0^+$, the biological relevant domain. The Jacobian Dh evaluated along $(s, c) \in S$ is given by

$$Dh|_S = NDf|_S \begin{pmatrix} -\dfrac{\alpha}{s + \alpha} & s + \alpha \\ \beta\dfrac{\alpha}{s + \alpha} & -\beta(s + \alpha) \end{pmatrix}$$

which has one trivial zero eigenvalue and $\lambda_1 = DfN|_S = -(\alpha+\beta(s+\alpha)^2)/(s+\alpha) < 0$, $\forall s \in \mathbb{R}_0^+$, as its nontrivial eigenvalue. Hence by Definition 3.6, $S = S_h = S_a$ is an attracting normally hyperbolic manifold.

Reduced Problem:
Here, DfN is a scalar and its value along S corresponds to the single nontrivial eigenvalue λ_1 of the layer problem. We obtain the following reduced problem (3.19),

$$\begin{pmatrix} \dot{s} \\ \dot{c} \end{pmatrix} = \left(\begin{pmatrix} 1 & 0 \\ 0 & 1 \end{pmatrix} + (1 - c + \beta(s + \alpha)) \begin{pmatrix} -(1-c) & s+\alpha \\ \beta(1-c) & -\beta(s+\alpha) \end{pmatrix} \right) \begin{pmatrix} 0 \\ -c \end{pmatrix}$$

which has to be evaluated along the critical manifold S. Since S is a graph over the s-coordinate chart, $c_0(s) = s/(s+\alpha)$, we can study the reduced flow in this single coordinate chart which leads to

$$\dot{s} = -\frac{s(s+\alpha)}{\alpha + \beta(s+\alpha)^2} \tag{3.32}$$

and confirms (2.18). It has $s = 0$ as an attracting equilibrium within the biochemical relevant domain $s \geq 0$.

Remark 3.16 *Note that no formal power series expansion of the slow manifold S_ε is needed to obtain the reduced problem (3.32); compare with the classic approach shown in Chap. 2 that uses such a power series ansatz. We clarify this observation in the following section.*

3.5 Slow Manifold Expansion and the Projection Operator

In general, the projection operator (3.20) provides the means to obtain a formula for the first order correction of S_ε for a general singularly perturbed system (3.15). Assume without loss of generality that the slow manifold S_ε is locally a graph

$$y = h(x, \varepsilon) = h_0(x) + \varepsilon h_1(x) + \ldots, \tag{3.33}$$

i.e., we assume that $D_y f|_S$ has locally full rank and $y = h_0(x)$ represents the corresponding critical manifold S, i.e., $f(x, h_0(x)) = 0$. Invariance under the flow (3.15) demands

$$\begin{aligned} \varepsilon \dot{y} &= (D_x h_0 + \varepsilon D_x h_1 + \ldots)\varepsilon \dot{x} \\ &= (D_x h_0 + \varepsilon D_x h_1 + \ldots)(N^x f + \varepsilon G^x) \\ &= (N^y f + \varepsilon G^y), \end{aligned} \tag{3.34}$$

where $z = (x, y)^\top$, $N = (N^x, N^y)^\top$ and $G = (G^x, G^y)^\top$. Evaluating this system by powers of ϵ gives the following system for the first order correction $O(\varepsilon)$:

$$(N^y - D_x h_0\, N^x) D_y f h_1(x) = -(G^y - D_x h_0\, G^x) \tag{3.35}$$

which defines $h_1(x)$ implicitly as a function of $h_0(x)$, i.e., the vectors G, N and the matrix $D_y f$ have to be evaluated at $y = h_0(x)$. Note that $(N^y - D_x h_0\, N^x) = DfN$ and $(G^y - D_x h_0\, G^x) = DfG$, where $Df = (-D_x h_0, I_{n-k})$ since $f(x, y) \approx y - h_0(x) = 0$ by definition of S. If we assume that the square matrix DfN is a regular matrix, i.e., $S = S_h$ or $S = S_n$ (no zero eigenvalues), then we obtain an explicit formula for the first order correction term:

$$h_1(x) = -(D_y f)^{-1}(DfN)^{-1}DfG. \tag{3.36}$$

Finally, plugging (3.33) into (3.15) and taking the limit $\varepsilon \to 0$ gives the desired reduced problem (3.19).

Remark 3.17 *If we calculate the first order correction term $c_1(s)$ for the enzyme kinetics model (2.3) by using formula (3.36), we replicate (2.17), i.e., $c_1(s) = -(D_c f)^{-1}(DfN)^{-1}DfG = -s/(\alpha + \beta(s + \alpha)^2)$.*

3.6 Comparison with the Standard Case

A singularly perturbed system in standard form (1.1) is equivalent to (3.10) with

$$z = \begin{pmatrix} x \\ y \end{pmatrix}, \quad N(z) = N = \begin{pmatrix} \mathbb{O}_{k,n-k} \\ \mathbb{I}_{n-k} \end{pmatrix}, \quad G(z) = \begin{pmatrix} g(x,y,\varepsilon) \\ \tilde{f}(x,y,\varepsilon) \end{pmatrix}. \tag{3.37}$$

Layer Problem:
The standard $(n - k)$-dimensional layer problem (3.4) is given by

$$\begin{pmatrix} x' \\ y' \end{pmatrix} = \begin{pmatrix} \mathbb{O}_k \\ f(x,y,0) \end{pmatrix}, \tag{3.38}$$

where $x \in \mathbb{R}^k$ plays the role of a parameter that defines each fast fibre $\mathcal{L}^c = \mathcal{L}^x$. Hence the layer problem is restricted globally to the $(n - k)$-dimensional subspace spanned by the fixed column vectors of $N(z) = N$ (the y-subspace) at each base point $(x, y) \in S$. Thus we identify $y \in \mathbb{R}^{n-k}$ as fast variables and $x \in \mathbb{R}^k$ as slow variables. This is in contrast to the more general $(n - k)$-dimensional layer problem (3.9) where such an identification is, in general, not given. Note, the nontrivial eigenvalues are the eigenvalues of the $(n-k)$-dimensional square matrix $DfN = D_y f$, i.e., the Jacobian with respect to the fast variable $y \in \mathbb{R}^{n-k}$.

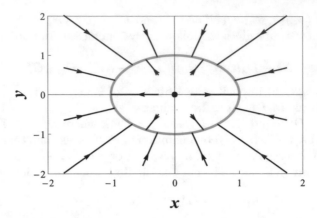

Fig. 3.3 Layer problem of a general singularly perturbed system (3.10) with functions defined in (3.41): the critical manifold S is a circle which is normally hyperbolic; the fibres are radial

Reduced Problem:
For $z \in S$, the standard k-dimensional reduced problem (3.19) is given by

$$\begin{pmatrix} \dot{x} \\ \dot{y} \end{pmatrix} = \begin{pmatrix} \mathbb{I}_k & \mathbb{O}_{k,n-k} \\ -(D_y f)^{-1} D_x f & \mathbb{O}_{n-k,n-k} \end{pmatrix} \begin{pmatrix} g \\ \tilde{f} \end{pmatrix} = \begin{pmatrix} g(x,y,0) \\ -(D_y f)^{-1} D_x f \, g(x,y,0) \end{pmatrix}.$$
(3.39)

In the case of a normally hyperbolic manifold, the critical manifold $S = S_h$ is a graph $y = Y_0(x)$ over the slow variable base $x \in \mathbb{R}^k$. The reduced problem in this coordinate chart is simply given by

$$\dot{x} = g(x, Y_0(x), 0).$$
(3.40)

In the standard case, isolated singularities on the critical manifold $S = S_h$ are defined by the k independent equations $g(x, Y_0(x), 0) = \mathbb{O}_k$ which is equivalent to Assumption 3.4.

Remark 3.18 *In a singularly perturbed system in standard form (1.1), a normally hyperbolic critical manifold S has always a (global) graph representation over the single slow coordinate chart x. This is not necessarily true for the general case as the following example shows. Take system (3.10) with*

$$z = \begin{pmatrix} x \\ y \end{pmatrix} \in \mathbb{R}^2, \ f(x,y) = x^2 + y^2 - 1, \ N(x,y) = -\begin{pmatrix} x \\ y \end{pmatrix}.$$
(3.41)

Here, the one-dimensional critical manifold S is the unit circle. It is an attracting critical manifold since its nontrivial eigenvalue $\lambda_1 = DfN|_S = -2 < 0$, $\forall (x,y) \in S$. Clearly, S cannot be represented in a single coordinate chart; see Fig. 3.3.

3.7 Local Transformation to Standard Form

As highlighted by Fenichel [30], the standard form (1.1) of a singularly perturbed system depends upon a special choice of local coordinates.

Lemma 3.5 *Given a singularly perturbed system in general form (3.10) with k-dimensional critical manifold S and $(n-k)$-dimensional regular level sets $\mathcal{L}^c$ (3.8) in a tubular neighbourhood $\mathcal{B} \subset \mathbb{R}^n$ of this critical manifold S, i.e., $\operatorname{rk} DL|_{\mathcal{B}} = k$, that form a local invariant nonlinear 'foliation' $\mathcal{L} = \cup_{c \in \mathcal{U}} \mathcal{L}^c$ of S for the corresponding layer problem. Then there exists a local coordinate system in $\mathcal{B}$ in which (3.10) takes the standard form (1.1).*

Proof We straighten locally the invariant nonlinear foliation $\mathcal{L}$ of the layer problem (3.9) as follows[6]: define local coordinates $z = (x, y)^\top$, $x \in \mathbb{R}^k$ and $y \in \mathbb{R}^{(n-k)}$, $G = (G^x, G^y)^\top$ and $N = (N^x, N^y)^\top$ with $k \times (n-k)$ matrix N^x such that the $(n-k) \times (n-k)$ matrix N^y is regular. By definition of the local invariant nonlinear foliation $\mathcal{L}$, $DL(z)N(z) = \mathbb{O}_{k,n-k}$ has to hold for any base point $z \in S$. The regularity of the $(n-k) \times (n-k)$ matrix N^y implies that $D_x L$ is locally a regular $k \times k$ matrix. It follows from the implicit function theorem that the level sets $\mathcal{L}^c$, i.e., $L(x, y) = c$, are locally given as graphs $x = X^c(y)$ over the y-coordinate chart. Hence we can define a local (nonlinear) coordinate transformation,

$$u = L(x, y), \qquad u \in \mathbb{R}^k, \tag{3.42}$$

with $x = M(u, y)$ as its inverse which transforms (3.10) locally to

$$\begin{pmatrix} u' \\ y' \end{pmatrix} = \begin{pmatrix} \mathbb{O}_k \\ N^y(M(u,y),y) \end{pmatrix} f(M(u,y),y) + \varepsilon \begin{pmatrix} DL\, G(M(u,y),y,\varepsilon) \\ G^y(M(u,y),y,\varepsilon) \end{pmatrix} = \begin{pmatrix} \varepsilon \tilde{g}(u,y,\varepsilon) \\ \tilde{f}(u,y,\varepsilon) \end{pmatrix}. \tag{3.43}$$

Note that the Jacobian of the corresponding layer problem evaluated along S is given by

$$\begin{aligned} J &= \begin{pmatrix} \mathbb{O}_{k,k} & \mathbb{O}_{k,n-k} \\ N^y D_x f D_u M & N^y (D_x f D_y M + D_y f) \end{pmatrix} \\ &= \begin{pmatrix} \mathbb{O}_{k,k} & \mathbb{O}_{k,n-k} \\ N^y D_x f (D_x L)^{-1} & N^y (Df N)(N^y)^{-1} \end{pmatrix}, \end{aligned} \tag{3.44}$$

where we use the fact that $D_u M = (D_x L)^{-1}$ and $D_y M = -(D_x L)^{-1} D_y L = N^x(N^y)^{-1}$ which follows from $u = L(M(u,y),y)$ and $DL(x,y)N(x,y) = \mathbb{O}_{k,n-k}$ for $(x,y) \in S$. This confirms Lemma 3.3, i.e., the nontrivial eigenvalues are the eigenvalues of the $(n-k) \times (n-k)$ matrix $Df N$. $\qquad\square$

---

[6] Formally we will have to introduce a sufficiently smooth local *cutoff function* for the vector field (3.9), respectively, (3.10) that takes only the neighbourhood $\mathcal{B} \subset \mathbb{R}^n$ of S into account.

Remark 3.19 *Due to Assumption 3.3, Lemma 3.5 holds in general, i.e., we can also straighten locally the 'foliation' $\mathcal{L}$ near lower dimensional subsets $S\backslash S_n$ of the critical manifold where the corresponding nonlinear 'fibres' have a tangency with $S\backslash S_n$; see also Remark 3.7.*

Lemma 3.6 *Given a singularly perturbed system in general form (3.10) with critical manifold S_n, i.e., all nontrivial eigenvalues are nonzero. Then there exists a local coordinate system in which (3.10) takes the standard form (1.1) with the set $\{y = 0\}$ as the local critical manifold S_n.*

Proof As shown in the proof of Lemma 3.5, the local transformation (3.42) that leads to system (3.43) straightens the local fast fibres $\mathcal{L}^c$ to the sets $\{u = c \in \mathbb{R}^k\}$. In a second step, we straighten locally the critical manifold S normal to the fast fibre bundle $\mathcal{L}$. Here we use the fact that we are dealing with a critical manifold S_n which guarantees a unique splitting of the phase space (3.5), i.e., S_n is transverse to the fibre bundle. This implies that the critical manifold is locally a graph over u-space in (3.43). Thus we define a local coordinate transformation

$$v = f(M(u,y),y), \tag{3.45}$$

with inverse $y = K(u,v)$ which transforms (3.43) locally to

$$\begin{pmatrix} u' \\ v' \end{pmatrix} = \begin{pmatrix} \mathbb{O}_k \\ DfN \end{pmatrix} v + \varepsilon \begin{pmatrix} DL\,G \\ Df\,G \end{pmatrix} = \begin{pmatrix} \varepsilon \tilde{g}(u,v,\varepsilon) \\ \tilde{f}(u,v,\varepsilon) \end{pmatrix}, \tag{3.46}$$

where $DfN = D_x fN^x + D_y fN^y$, $DL\,G = D_x LG^x + D_y LG^y$ and $Df\,G = D_x fG^x + D_y fG^y$. $\square$

Remark 3.20 *System (3.46) reflects a simple geometric structure where the manifold S and the nonlinear fibre bundle $\mathcal{L}$ are locally straightened. This system can be simplified even further into a local Fenichel normal form where individual stable and unstable fibres are straightened locally as well; see, e.g., [30, 46, 61] for details.*

It is worthwhile to mention that Fenichel [30] straightened the critical manifold S first and then its (local) fibres. The order does not matter in the case $S = S_h$ or in the case $S = S_n$, but it becomes important when studying singularly perturbed systems near $S\backslash S_n$ as described in Chap. 4, because one can only straighten (locally) the manifold S or the nonlinear 'foliation' $\mathcal{L}$, but not both.

Remark 3.21 *Recall from Lemma 3.2, the condition $DL(x,y)N(x,y)f(x,y) = \mathbb{O}_k$ defines the (vector-valued) function $L(x,y)$ which, in general, cannot be solved explicitly. Local representations of the critical manifold S and its fast nonlinear foliation $\mathcal{L}$ that would allow for local straightening of the fibres are, in general, hard to find[7]. Hence, in practice we want to avoid such lo-*

---

[7] It would be also tedious to find approximations of these fibres through formal power series expansions.

cal transformations to standard form and deal directly with the general form of the singularly perturbed problem (3.10), *which is the main focus of this manuscript.*

3.8 Normally Hyperbolic Results for $0 < \varepsilon \ll 1$

Due to local equivalence of systems (1.1) and (3.10) described in Lemma 3.5, Fenichel theory applies directly to the general system (3.10), respectively, (3.15) when dealing with a normally hyperbolic critical manifold $S_h \subseteq S$; see [30].

For completeness, we recall the main result of Fenichel theory here which guarantees the persistence of a normally hyperbolic manifold $S_{h,\varepsilon}$ close to $S_h \subseteq S$ and the persistence of a slow flow on $S_{h,\varepsilon}$ close to the reduced flow of S_h. It also guarantees the persistence of the local stable and unstable manifolds close to $W_{loc}^s(S_h)$ and $W_{loc}^u(S_h)$ and corresponding fast fibration (foliation).

Theorem 3.1 (Fenichel's Theorem, cf. [30, 46, 61])
Given system (3.10) *with* C^r *vector field* H ($1 < r < \infty$). *Suppose* $S_h \subseteq S$ *is a compact normally hyperbolic manifold, possibly with boundary. Then for* $\varepsilon > 0$ *sufficiently small the following holds:*

(i) There exists a C^r *smooth manifold* $S_{h,\varepsilon}$, *locally invariant under the flow* (3.10), *that is* C^r $O(\varepsilon)$ *close to* S_h.
(ii) The slow flow on $S_{h,\varepsilon}$ *converges to the reduced flow on* S_h *as* $\varepsilon \to 0$.
(iii) There exist C^r *smooth stable and unstable manifolds*

$$W_{loc}^s(S_{h,\varepsilon}) = \bigcup_{z \in S_{h,\varepsilon}} W_{loc}^s(z), \quad W_{loc}^u(S_{h,\varepsilon}) = \bigcup_{z \in S_{h,\varepsilon}} W_{loc}^u(z), \quad (3.47)$$

locally invariant under the flow (3.10), *that are* C^r $O(\varepsilon)$ *close to* $W_{loc}^s(S_h)$ *and* $W_{loc}^u(S_h)$, *respectively.*
(iv) The stable foliation $\{W_{loc}^s(z)|\, z \in S_{h,\varepsilon}\}$ *is* C^{r-1} *and (positively) invariant, i.e.,*

$$W_{loc}^s(z) \cdot t \subseteq W_{loc}^s(z \cdot t)$$

for all $t \geq 0$ *such that* $p_\varepsilon \cdot t \in S_{h,\varepsilon}$, *where* $\cdot t$ *denotes the solution operator of system* (3.10).
(v) The unstable foliation $\{W_{loc}^u(z)|\, z \in S_{h,\varepsilon}\}$ *is* C^{r-1} *and (negatively) invariant, i.e.,*

$$W_{loc}^u(z) \cdot t \subseteq W_{loc}^u(z \cdot t)$$

for all $t \leq 0$ *such that* $z \cdot t \in S_{h,\varepsilon}$, *where* $\cdot t$ *denotes the solution operator of system* (3.10).

Remark 3.22 $S_{h,\varepsilon}$ *is, in general, not unique but all representations of* $S_{h,\varepsilon}$ *lie exponentially close in* ε *from each other, i.e., all r-jets are uniquely determined.*

Remark 3.23 *We assume that a compact, simply connected, k-dimensional smooth manifold M with boundary* ∂M *implies that its boundary is a* $(k-1)$-*dimensional smooth manifold. A compact manifold with boundary is called* inflowing/overflowing invariant *if the vector field inside the manifold M is tangent to the manifold and along the boundary* ∂M *it points everywhere inward/outward. Both are special cases of a locally invariant manifold M where the vector field inside the manifold M is tangent to the manifold, but no special structure is prescribed along the boundary* ∂M.

Remark 3.24 *Fenichel's result for general normally hyperbolic manifolds [30] is equivalent to Tikhonov's result for attracting critical manifolds [104]. It is this result that provides the mathematical justification for a QSSR under the assumption that the corresponding critical manifold is normally hyperbolic and attracting; see Sect. 2.1. This implies that after some initial transient fast time, the flow on the attracting slow manifold is to leading order described by the reduced flow[8]. Goeke, Noethen and Walcher [33, 85] provide a comprehensive discussion on the general setup of GSPT explaining when QSSR is justified or when it leads to erroneous results.*

Remark 3.25 *We would like to point out that there is an important distinction to make whether we are in a normally hyperbolic regime* S_h *or in the neighbourhood of* S_n/S_h *where normal hyperbolicity is lost due to a pair of complex eigenvalues crossing the imaginary axis. While one can locally straighten the critical manifold and the fast fibres in both cases as outlined in Lemma 3.6, Fenichel theory only applies to the normally hyperbolic case* $S = S_h$. *The case of a complex eigenvalue pair crossing the imaginary axis leads to a well-known* delayed loss of stability *phenomenon. We will not discuss this case here, but refer the interested reader to the corresponding results presented in, e.g., [41, 84]; see also Sect. 7.1.*

---

[8] Provided that this reduced flow is robust, i.e., does not undergo any bifurcation(s) itself.

Chapter 4
Loss of Normal Hyperbolicity

This chapter extends the coordinate-independent GSPT beyond *Fenichel theory*. Loss of normal hyperbolicity is one essential ingredient for a singularly perturbed system to switch between slow and fast dynamics as observed in many relaxation oscillator models; see Chap. 2. Geometrically, loss of normal hyperbolicity occurs generically along (a union of) codimension-one submanifold(s) of S where a nontrivial eigenvalue of the layer problem crosses the imaginary axis. Within this set $S \backslash S_h$, we distinguish two subsets F and AH, where

- the set F is associated with the crossing of a real eigenvalue,
- the set AH is associated with the crossing of a pair of complex conjugate eigenvalues (with nonzero imaginary part).

Note that $F \subseteq S \backslash S_n$ and $AH \subseteq S_n \backslash S_h$. In this manuscript, we focus primarily on the real eigenvalue case associated with the set F. The imaginary eigenvalue case associated with the set AH will be briefly discussed in Chap. 7.

Remark 4.1 *The codimension-two Bogdanov–Takens case has been treated in [31].*

4.1 Layer Flow and Contact Points

As shown in Lemma 3.3, the $(n-k)$-dimensional square matrix DfN encodes the nontrivial eigenvalues of the n-dimensional Jacobian NDf of the layer problem (3.9) along S. This $(n-k)$-dimensional square matrix DfN becomes singular if one of the column vectors $N^i(z)$, $i = 1, \ldots n - k$, or a linear combination of these vectors, aligns with the tangent space $T_z S$, i.e., when a nontrivial (real) eigenvalue of the layer problem becomes zero.

© Springer Nature Switzerland AG 2020
M. Wechselberger, *Geometric Singular Perturbation Theory Beyond the Standard Form*, Frontiers in Applied Dynamical Systems: Reviews and Tutorials 4, https://doi.org/10.1007/978-3-030-36399-4_4

Definition 4.1 *The set*

$$F := \{z \in S \;:\; \mathrm{rk}(DfN) = n - k - 1\} \subseteq S \backslash S_n \qquad (4.1)$$

denotes the (union of) codimension-one submanifold(s) along which exactly one nontrivial eigenvalue vanishes.

A necessary condition for this (not necessarily connected) subset $F \subseteq S \backslash S_n$ is given by

$$\det(DfN)|_F = 0. \qquad (4.2)$$

Along this set F, the algebraic multiplicity of the zero eigenvalue of the n-dimensional Jacobian NDf of the layer problem (3.9) is $k + 1$ while the geometric multiplicity is still k reflecting the dimension of the critical manifold S which by Assumption 3.2 is the zero level set of the function h with constant rank $n - k$ or equivalently by (3.6) is a regular zero level set of the function f, i.e., the $(n - k) \times n$ matrix $Df|_S$ has full (row) rank $n - k$. From condition (4.2) we can deduce the linear combination of the vectors $N^i(z)$, $i = 1, \ldots n - k$, which aligns with the tangent space $T_z S$ at $z \in F$. It is given by a nontrivial column vector of the rank one matrix

$$N \, \mathrm{adj}\,(DfN)|_F \neq \mathbb{O}_{n,n-k}, \qquad (4.3)$$

where $\mathrm{adj}\,(DfN)$ is the *adjoint*[1] of the $(n - k)$-dimensional square matrix DfN.

Remark 4.2 *The adjoint of a $(n-k)$-dimensional square matrix is the transpose of its co-factor matrix, and these matrices are related through*

$$adj\,(DfN)\,(DfN) = (DfN)\,adj\,(DfN) = \det(DfN)\,\mathbb{I}_{n-k}. \qquad (4.4)$$

The matrix $adj\,(DfN)$ has full rank for $z \in S_n$ since $\det(DfN) \neq 0$ for all $z \in S_n$. On the other hand, the matrix $adj\,(DfN)$ has rank one for $z \in F$ since $adj\,(DfN)(DfN) = (DfN)adj\,(DfN) = \mathbb{O}_{n-k,n-k}$, for all $z \in F$ and the $(n - k)$-dimensional square matrix DfN has rank $n - k - 1$ by Definition 4.1. Thus $adj\,(DfN)$ is still a well-defined nontrivial matrix for $z \in F$, albeit of rank one only, and it encodes the left and right nullspaces of the matrix DfN. In the case $k = n - 1$, $adj\,(Df\,N) := \mathbb{I}_1 = 1$ by definition.

Thus a nontrivial column vector of (4.3) forms a right nullvector of the n-dimensional Jacobian NDf of the layer problem (3.9) along F. Similarly, a nontrivial row vector of the rank one matrix

$$\mathrm{adj}\,(DfN)Df|_F \neq \mathbb{O}_{n-k,n} \qquad (4.5)$$

represents a left nullvector of the n-dimensional Jacobian NDf of the layer problem (3.9) along F.

---

[1] This matrix is also referred to as the *adjugate*.

Definition 4.2 *The set F forms a set of* contact points *between the critical manifold S and the layer flow (3.9). At a contact point $z \in F$, the layer flow (3.9) has contact of order $l \geq 1$ with the critical manifold S, if there exists a one-dimensional (sub-)manifold $C^z \subseteq S$ and a one-dimensional nonlinear fast 'fibre' $\mathcal{L}^z$ in the corresponding $(k+1)$-dimensional local centre manifold of the layer problem near $z \in F$ and a local coordinate chart containing $z \in F$ such that both one-dimensional regular submanifolds are graphs of smooth functions, $p(v)$ for C^z and $c(v)$ for $\mathcal{L}^z$, $v \in [-v_0, v_0]$, which are l-th order equivalent at $c(0) = p(0) = z \in F$, i.e., $c^{(m)}(0) = p^{(m)}(0)$ for $m = 1, \ldots l$, and $c^{(l+1)}(0) \neq p^{(l+1)}(0)$.*

Remark 4.3 *This is a coordinate-independent definition; see, e.g., [87].*

4.1.1 Contact of Order One and Its Comparison to the Standard Case

Our aim is to derive a computable criterion for Definition 4.2 with $l = 1$ to show contact of order one at a point $z \in F$ between S and the flow of a general layer problem (3.9). There are two main approaches to achieve this goal: either straighten (locally) the critical manifold S or straighten (locally) the corresponding fast nonlinear foliation $\mathcal{L}$. We follow the latter approach since this allows us to compare the criterion for contact of order one with that for a standard layer problem (3.38).

Lemma 4.1 *In a singularly perturbed system of the general form (3.10), the critical manifold S has contact of order one with the layer flow (3.9) at a contact point $z \in F$ if*

$$\operatorname{rk} Df|_F = n - k,$$
$$l \cdot (D^2 f(Nr, Nr) + Df DN(Nr, r))|_F \neq 0, \tag{4.6}$$

where l, respectively, r is a nontrivial row, respectively, column vector of the matrix $\operatorname{adj}(Df\,N)|_F$, and $D^2 f(Nr, Nr)$ and $Df DN(Nr, r)$ are bilinear forms evaluated at a contact point $z \in F$ and given in component-wise notation by

$$[D^2 f]_i(Nr, Nr) = \sum_{l,m=1}^{n} \sum_{j,s=1}^{n-k} \frac{\partial^2 f_i}{\partial z_l \partial z_m}(N_{mj} r_j)(N_{ls} r_s), \quad i = 1, \ldots, n-k,$$

$$[Df DN]_i(Nr, r) = \sum_{l,m=1}^{n} \sum_{j,s=1}^{n-k} \frac{\partial f_i}{\partial z_l} \frac{\partial N_{lj}}{\partial z_m}(N_{ms} r_s) r_j, \quad i = 1, \ldots, n-k,$$

$$\tag{4.7}$$

Proof The first condition necessarily reflects the fact that the critical manifold S is a regular zero level set of the function f, i.e., $Df|_S$ has constant rank $(n - k)$ everywhere including F.

Recall from Lemma 3.5 that there exists a local coordinate system $(u, y) \in \mathbb{R}^k \times \mathbb{R}^{n-k}$ with $u = L(x, y)$ and its inverse $x = M(u, y)$ in which a general singularly perturbed system (3.10) takes the standard form (3.43). We would like to stress the fact that this Lemma also applies in the case of loss of normal hyperbolicity; see also Remark 3.19. The corresponding layer problem in this local coordinate system is given by

$$\begin{pmatrix} u' \\ y' \end{pmatrix} = \begin{pmatrix} \mathbb{O}_{k,n-k} \\ N^y(M(u,y),y) \end{pmatrix} f(M(u,y),y) = \begin{pmatrix} \mathbb{O}_k \\ \tilde{f}(u,y), \end{pmatrix} \qquad (4.8)$$

where $u \in \mathbb{R}^k$ represents the local slow variable and $y \in \mathbb{R}^{n-k}$ represents the local fast variable. Loss of normal hyperbolicity of the k-dimensional critical manifold S happens along a $(k-1)$-dimensional set F where the $(n-k)$-dimensional square matrix

$$D_y \tilde{f} = N^y Df N (N^y)^{-1}$$

of the Jacobian (3.44) has a single zero eigenvalue, i.e., the $(n-k)$-dimensional square matrix DfN is singular along F and $\mathrm{rk}(DfN) = n - k - 1$, which follows directly from (4.1).

Note that a nontrivial row vector of the rank one matrix $\mathrm{adj}\,(DfN)(N^y)^{-1}|_F$ encodes a left nullvector $\tilde{l}$ of $D_y\tilde{f}|_F$ and a nontrivial column vector of the rank one matrix $N^y \mathrm{adj}\,(DfN)|_F$ encodes a right null-vector $\tilde{r}$ of $D_y\tilde{f}|_F$. The contact between S and the layer flow along F happens in the direction of the right nullvector $\tilde{r}$, i.e., the nontrivial centre direction of the layer problem. We isolate this direction of contact through the linear coordinate transformation

$$\begin{pmatrix} v \\ w \end{pmatrix} = \begin{pmatrix} \tilde{l} \\ \tilde{Q} \end{pmatrix} y, \qquad (4.9)$$

with corresponding inverse transformation

$$y = \begin{pmatrix} \tilde{r} & \tilde{P} \end{pmatrix} \begin{pmatrix} v \\ w \end{pmatrix} = \tilde{r}v + \tilde{P}w, \qquad (4.10)$$

where $v \in \mathbb{R}$, $w \in \mathbb{R}^{n-k-1}$, $\tilde{l} = l(N^y)^{-1}$ and $\tilde{r} = N^y r$ denote the aforementioned left and right null vectors of the Jacobian $D_y\tilde{f}|_F$, $\tilde{Q} = Q(N^y)^{-1}$ is a corresponding $(n - k - 1) \times (n - k)$ matrix and $\tilde{P} = N^y P$ is a corresponding $(n - k) \times (n - k - 1)$ matrix such that

$$\begin{pmatrix} \tilde{r} & \tilde{P} \end{pmatrix} \begin{pmatrix} \tilde{l} \\ \tilde{Q} \end{pmatrix} = \begin{pmatrix} \tilde{l} \\ \tilde{Q} \end{pmatrix} \begin{pmatrix} \tilde{r} & \tilde{P} \end{pmatrix} = \mathbb{I}_{n-k}, \qquad (4.11)$$

which leads to the transformed system

$$
\begin{pmatrix} u' \\ v' \\ w' \end{pmatrix} = \begin{pmatrix} \mathbb{O}_{k,n-k} \\ \tilde{l} \\ \tilde{Q} \end{pmatrix} \tilde{f}(u, \tilde{r}v + \tilde{P}w) = \begin{pmatrix} \mathbb{O}_{k,n-k} \\ l \\ Q \end{pmatrix} f(M(u, \tilde{r}v + \tilde{P}w), \tilde{r}v + \tilde{P}w).
$$

$$(4.12)$$

The corresponding Jacobian evaluated along S is given by

$$
\tilde{J}|_S = \begin{pmatrix} \mathbb{O}_{k,k} & \mathbb{O}_{k,1} & \mathbb{O}_{k,n-k-1} \\ lD_x f(D_x L)^{-1} & lDfNr & lDfNP \\ QD_x f(D_x L)^{-1} & QDfNr & Q\,DfN\,P \end{pmatrix},
$$

$$(4.13)$$

where we used the fact that $D_u M = (D_x L)^{-1}$. This Jacobian evaluated along $F \subset S$ is then given by

$$
\tilde{J}|_F = \begin{pmatrix} \mathbb{O}_{k,k} & \mathbb{O}_{k,1} & \mathbb{O}_{k,n-k-1} \\ lD_x f(D_x L)^{-1} & 0 & \mathbb{O}_{1,n-k-1} \\ QD_x f(D_x L)^{-1} & \mathbb{O}_{n-k-1,1} & Q\,DfN\,P \end{pmatrix},
$$

$$(4.14)$$

which confirms that the contact between the manifold S and the straight fast fibres along F happens in the (nontrivial) centre direction given by the v-coordinate axis in system (4.12). Based on Assumption 3.2, a necessary condition for the local existence of the k-dimensional manifold S near F is that $\tilde{J}$ has rank $n - k$ along S including F, i.e., rk $\tilde{J}|_F = n - k$, which implies

$$
l \cdot D_x f|_F \neq \mathbb{O}_k,
$$

$$(4.15)$$

and this is a necessary condition for the first condition in (4.6) to hold.

To derive the second condition in (4.6), we recall from Definition 4.2 that first order contact means that the curvature of the layer flow at a contact point $z \in F$ in the nontrivial centre direction is not identical to that of S. For the 'standard' system (4.12) where the fast fibres are straight, this implies that the critical manifold S has to be locally parabolic near $z \in F$ over the coordinate chart $v \in \mathbb{R}$ associated with the nontrivial centre direction. Thus expanding the right-hand side of the v' equation in (4.12) at such a contact point $z \in F$ gives

$$
\begin{aligned}
v' &= \tilde{l} \cdot \tilde{f}(u, \tilde{r}v + \tilde{P}w) \\
&= l \cdot f(M(u, \tilde{r}v + \tilde{P}w), \tilde{r}v + \tilde{P}w) \\
&= l \cdot D_x f(D_x L)^{-1} u + l \cdot (D^2 f(Nr, Nr) + DfDN(Nr, r))v^2 + \dots,
\end{aligned}
$$

$$(4.16)$$

where $D^2 f(Nr, Nr)$ and $DfDN(Nr, r)$ are bilinear forms given in component-wise notation by (4.7). All partial derivatives of the (functional) components of the vector f and the matrix N as well as the matrix N itself are evaluated at the corresponding contact point $z \in F$ that defines the left, respectively, right nullvector, l, respectively, r, of the $(n-k)$-dimensional square matrix DfN. A non-vanishing (scalar) coefficient of v^2 in (4.16), i.e.,

$$l \cdot (D^2 f(Nr, Nr) + DfDN(Nr, r)) \neq 0,$$

replicates the second condition in (4.6). Both conditions in (4.6) show that S is locally parabolic over the v-coordinate chart in the direction of the non-vanishing u component(s) and, hence, there is contact of order one at $z \in F$ between the layer flow and the critical manifold S. $\qquad\square$

This proof highlights that a first order contact in a singularly perturbed system in standard form (1.1) describes a *folded* critical manifold S where the codimension-one submanifold F is usually referred to as a *fold* to reflect its geometric property viewed over the slow coordinate chart $x \in \mathbb{R}^k$. This fold F corresponds to a codimension-one *saddle-node* bifurcation in the standard layer problem under the variation of the parameter (slow variable) $x \in \mathbb{R}^k$ along a suitable path $x(s)$, $s \in I$; see, e.g., Kuznetsov [63]:

Definition 4.3 *The k-dimensional critical manifold S of a standard singularly perturbed system (1.1) is (locally) folded if there exists a subset $F \subset S$ that forms a $(k-1)$-dimensional manifold defined by*

$$F := \{(x,y) \in S \mid \mathrm{rk}(D_y f) = n - k - 1; \ l \cdot D_x f \neq \mathbb{O}_k, \ l \cdot D_{yy}^2 f(r,r) \neq 0\}, \tag{4.17}$$

where l and r denote the left and right null vector of the Jacobian $D_y f$, respectively, and $D_{yy}^2 f(r,r)$ is a bilinear form given in component-wise notation by

$$[D_{yy}^2 f]_i(r,r) = \sum_{j,l=1}^{n-k} \left. \frac{\partial^2 f_i}{\partial y_j \partial y_l} \right|_{(x,y) \in F} r_j r_l, \quad i = 1, \ldots, n-k. \tag{4.18}$$

Note that the three fold conditions in (4.17) are equivalent to the three contact of order one conditions in (4.1) and (4.6), since $DfN = D_y f$ and $N = (N^x, N^y) = (\mathbb{O}_{k, n-k}, \mathbb{I}_{n-k})$.

Lemma 4.2 *Let the set F consist entirely of order one contact points. Then the determinant $\det(DfN)|_S$ changes sign as one crosses the set F from one branch of S to its adjacent.*

Proof By definition, we have $\det(DfN) = 0$ for all $z \in F$ which reflects the existence of a single nontrivial zero eigenvalue λ_1 of the matrix DfN. This single nontrivial eigenvalue has been isolated in system (4.12), as can be seen from the block structure of the corresponding Jacobian (4.14) evaluated at $z \in F$. This block structure of the Jacobian for $z \in F$ implies that this critical nontrivial eigenvalue λ_1 is given to leading order by $\lambda_1 \simeq l DfNr$ for any point $z \in S$ in a neighbourhood of F; see (4.13). All other nontrivial eigenvalues of the matrix DfN are encoded to leading order in the $(n-k-1)$-dimensional square matrix $QDfNP$, and they are bounded away from zero (and the imaginary axis) in a neighbourhood of F, i.e., we have a spectral gap. Hence, a sign change of $\det(DfN)|_S$ in a neighbourhood of F requires a sign change of λ_1.

Take a path c along S with local parameterisation $c(s) = (u_c(s), v_c(s), w_c(s))$, $s \in [-s_0, s_0]$ such that $c(0) = z \in F$, i.e., $\lambda_1(c(0)) = 0$, which coincides with the direction of the contact of at $z \in F$ in system (4.12), i.e., $\dot{c}(0) = (\mathbb{O}_k, 1, \mathbb{O}_{n-k-1})$; compare with Definition 4.2. Then

$$\frac{d}{ds}\lambda_1(c(s))|_{s=0} \simeq (l\, D(Df\, Nr)\dot{c}(s))|_{s=0} = l\cdot(D^2 f(Nr, Nr) + Df\, DN(Nr, r)) \neq 0\,,$$

by Lemma 4.1. By continuity, this transversality result will hold for any path nearby crossing F transversally which indicates that λ_1 changes its sign as we cross F.   □

Remark 4.4 *The order one contact condition (4.6) in Lemma 4.1 is thus a condition on the sign change of the determinant DfN near F and, hence, indicates different stability properties of adjacent normally hyperbolic branches of S near F.*

4.2 Reduced Flow Near Contact Points

Recall that the projection operator Π^S (3.20) is only well defined if the $(n-k)$-dimensional square matrix DfN is non-singular. This is the case for all $z \in S_n$ but DfN is singular along contact points $z \in F \subseteq S\backslash S_n$. Based on Definition 4.1, we focus on the case of a single zero eigenvalue, i.e., the rank of the matrix DfN along $z \in F$ is $n-k-1$. To uncover the properties of the reduced problem (3.19) in a neighbourhood of F, we replace the inverse $(DfN)^{-1}$ through the relationship with its adjoint defined in (4.4), i.e., $(Df\, N)^{-1} = \text{adj}\,(DfN)/\det(DfN)$, which gives the following equivalent reduced problem

$$\dot{z} = \Pi^S G(z, 0) = \left(\mathbb{I}_n - \frac{1}{\det(DfN)} N\text{adj}\,(DfN)Df\right) G(z, 0)\,. \qquad (4.19)$$

While system (4.19) is still singular for $z \in F$ where $\det(Df\, N)$ vanishes, we can (formally) remove this singularity in (4.19) by the time transformation $d\tau = -det(Df\, N)d\tau_1$ to obtain the corresponding *desingularised problem*

$$\dot{z} = \Pi^S_d G(z, 0) = (-\det(DfN)\mathbb{I}_n + N\text{adj}\,(DfN)Df)\, G(z, 0), \qquad (4.20)$$

where Π_d^S denotes the desingularised 'projection' operator[2] and, with a slight abuse of notation, the overdot denotes now differentiation with respect to τ_1. This desingularised problem can be studied for all $z \in S$ in a neighbourhood of $F \subset S$. For $z \in S_n$ and $\det(DfN) < 0$, the reduced flow (4.19) is equivalent to (4.20). For $z \in S_n$ and $\det(DfN) > 0$, the flow direction of (4.20) has to be reversed to obtain the reduced flow (4.19). Thus up to a possible orientation change in the flow direction, the reduced and the desingularised problem are equivalent. Thus we can (formally)[3] study the reduced flow via the desingularised problem (4.20) in a neighbourhood of F keeping in mind in which subsets of S the flow direction of (4.20) has to be reversed.

Remark 4.5 *In the generic case of order one contact along F, Lemma 4.2 shows that there is a sign change of $\det(DfN)$ as F is crossed, i.e., the flow direction has to be reversed on one of the two branches of S adjacent F.*

We note that a singularity of the reduced problem (4.19) away from F is also a singularity of the desingularised problem (4.20), i.e., a point $z_{eq} \in S\backslash F$, where

$$(-\det(DfN)\mathbb{I}_n + N\mathrm{adj}\,(DfN)Df)\,G(z,0) = \mathbb{O}_n \qquad (4.21)$$

is an equilibrium for both systems. In general, we expect such an equilibrium of (4.20) to be isolated under Assumption 3.4, i.e., $\Pi_d^S G(z,0) = 0$ defined in (4.21) represents k independent equations for $z \in S\backslash F$, and there are two possibilities for such an isolated equilibrium to occur: either $G(z_{eq},0) = 0$ or $\Pi_d^S G(z_{eq},0) = 0$ where $G(z_{eq},0) \neq 0$; compare with the discussion below Assumption 3.4.

Definition 4.4 *A contact point $z \in F$ where*

$$N\,adj\,(DfN)Df\,G(z,0) \neq \mathbb{O}_n \qquad (4.22)$$

is called a regular jump point.

Solutions of the reduced problem (4.19) in a neighbourhood of F approach regular jump points (4.22) in forward or backward time and they cease to exist at regular jump points due to a finite time blow-up, i.e., the reduced problem (4.19) has a pole at jump points. This finite time blow-up of the reduced flow at a jump point happens exactly in the eigendirection of the defect of the layer problem encoded in the matrix $\mathrm{adj}\,(DfN)$, i.e., the projector $\Pi_d^S|_F = N\mathrm{adj}\,(DfN)Df$ for $z \in F$ has rank one[4] and the corresponding nontrivial column vector of this projector encodes this unique direction of

---

[2] We have $(\Pi_d^S)^2 = -\det(DfN)\Pi_d^S$, i.e., this is only a 'projection' operator up to the time rescaling factor $-\det(DfN)$ which does not cause any troubles from a dynamical systems point of view.

[3] A blow-up analysis near contact points justifies this formal approach.

[4] For $z \in S_h$, the projectors Π^S and Π_d^S are equivalent and have the same rank which is $k \geq 1$.

finite time blow-up for the reduced problem (4.19). This alignment of the reduced flow and the layer flow at a jump point $z \in F$ defines the locus (and direction) where a switch from slow to fast dynamics (or vice versa) in the full system is possible; we state the corresponding results in Sect. 4.5.

4.2.1 Comparison with the Standard Case

The corresponding standard k-dimensional desingularised problem (4.20) is given by

$$\begin{pmatrix} \dot{x} \\ \dot{y} \end{pmatrix} = \begin{pmatrix} -\det(D_y f)\mathbb{I}_k & \mathbb{O}_{k,n-k} \\ \text{adj}\,(D_y f)D_x f & \mathbb{O}_{n-k,n-k} \end{pmatrix} \begin{pmatrix} g \\ \tilde{f} \end{pmatrix} = \begin{pmatrix} -\det(D_y f)g(x,y,0) \\ \text{adj}\,(D_y f)D_x f\, g(x,y,0) \end{pmatrix}.$$
(4.23)

Note that $\det(D_y f) = 0$ along F. Hence locally near F, the critical manifold has to be studied in a coordinate chart that includes at least one fast coordinate y_i, $1 \le i \le (n-k)$. The regular jump point condition (4.22) for a point $z \in F$ becomes

$$\text{adj}\,(D_y f)D_x f\, g(x,y,0) \neq \mathbb{O}_{n-k}.$$
(4.24)

Near jump points, solutions have the possibility to switch from the slow to the fast time scale or vice versa. The corresponding standard results which describe the flow near jump points for $\varepsilon \neq 0$ can be found in Sect. 4.5. These local results apply directly to the general setting through (local) equivalence.

Remark 4.6 *In the case of a normally hyperbolic manifold, the critical manifold is a graph $y = Y_0(x)$ over the slow variable base $x \in \mathbb{R}^k$ and the desingularised problem (4.23) in this slow coordinate chart is simply given by*

$$\dot{x} = -\det(D_y f)g(x, Y_0(x), 0),$$
(4.25)

where $\det(D_y f) \neq 0$. Compare with (3.40) which is the equivalent system up to the sign of $\det(D_y f)$.

4.3 Travelling Waves in ARD Models Revisited

Recall system (2.69), the travelling wave ODE of the ARD system (2.66), which is a singularly perturbed system in general form with

$$z = \begin{pmatrix} u \\ v \end{pmatrix} \in \mathbb{R}^2, \quad N(u,v) = \begin{pmatrix} 1 \\ u - c \end{pmatrix}, \quad f(u,v) = v,$$
(4.26)

and

$$G(u, v, \varepsilon) = \begin{pmatrix} 0 \\ u(1 - u) \end{pmatrix}. \tag{4.27}$$

The aim is to find heteroclinic connections between the asymptotic end states $(u_\pm, 0)$ of this system where $u_- = 1$ and $u_+ = 0$. Such connections correspond to travelling wave profiles of the original ARD model (2.66).

Layer Problem:
Taking the singular limit $\varepsilon \to 0$ in (2.69) gives the layer problem (2.70),

$$\begin{pmatrix} u' \\ v' \end{pmatrix} = N(u, v) f(u, v) = \begin{pmatrix} 1 \\ u - c \end{pmatrix} v. \tag{4.28}$$

The regular level set $v = 0$ (2.71) defines the one-dimensional critical manifold S of this problem. The Jacobian of the layer problem evaluated along $(u, v) \in S$ is given by

$$N Df|_S = \begin{pmatrix} 0 & 1 \\ 0 & u - c, \end{pmatrix}$$

which has one trivial zero eigenvalue and one nontrivial eigenvalue $\lambda_1 = Df N|_S = u - c$. Thus S loses normal hyperbolicity for $u = c > 0$ at the single contact point

$$F = \{(u, v) \in S \ : \ \det(Df N) = 0\} = \{(c, 0)^\top\}. \tag{4.29}$$

There is order one contact between the layer flow and the critical manifold at F since conditions (4.6) in Lemma 4.1 are fulfilled, i.e.,

$$\text{rk } Df|_F = n - k = 1,$$

$$l \cdot (D^2 f(Nr, Nr) + Df DN(Nr, r))|_F = Df DN N|_F = (0\ 1) \begin{pmatrix} 0 & 0 \\ 1 & 0 \end{pmatrix} \begin{pmatrix} 1 \\ 0 \end{pmatrix} = 1 \neq 0.$$

It follows that the critical manifold $S = S_a \cup F \cup S_r$ consists of an attracting branch S_a for $u < c$, a repelling branch for $u > c$.

Remark 4.7 *Recall, solutions of the layer problem of the ARD model evolve along parabolas $v = u^2/2 - cu + k$, (2.72), parameterised by k, i.e., this family of parabolas represents the fast fibration (foliation) of this layer problem. This confirms explicitly the order one contact of the fast fibre $v = (u - c)^2/2$ obtained for $k = c^2/2$ at $F = (c, 0)$ with the critical manifold S.*

Reduced Problem:
Here, we study the corresponding desingularised problem (4.20) given by

$$\begin{pmatrix} \dot{u} \\ \dot{v} \end{pmatrix} = \Pi_d^S G(u, v, 0) = \left(-(u - c) \begin{pmatrix} 1 & 0 \\ 0 & 1 \end{pmatrix} + \begin{pmatrix} 0 & 1 \\ 0 & u - c \end{pmatrix} \right) \begin{pmatrix} 0 \\ u(1 - u), \end{pmatrix}$$

which has to be evaluated along the critical manifold S given by $v_0(u) = 0$. Since S is a graph over the u-coordinate chart, we study the desingularised flow in this single coordinate chart and obtain

$$\dot{u} = u(1 - u); \tag{4.30}$$

compare with the corresponding reduced problem (2.77). Since $\lambda_1 = u - c$, the flow of this desingularised system has to be reversed on S_r, i.e., for $u > c$, to obtain the corresponding reduced flow. Note further that the single nontrivial condition (4.22) for the contact point F (4.29) to be a regular jump point, i.e.,

$$N \operatorname{adj}(DfN) Df\, G(u, v, 0)|_F = \begin{pmatrix} c(1 - c) \\ 0 \end{pmatrix} \neq \mathbb{0}_2 \tag{4.31}$$

is satisfied for $c \neq 1$. Hence, we have to distinguish two cases with respect to the regular jump point F: $0 < c < 1$ and $c > 1$.

As explained in Sect. 2.3 for $c > 1$, we identify smooth travelling wave profiles entirely on S_a described solely by the reduced problem that connect the two end states $u_\mp$; see Fig. 2.15. As shown in Chap. 3, Fenichel theory provides results on the existence of a nearby travelling wave located on $S_{a,\varepsilon}$ for sufficiently small perturbations $0 < \varepsilon \ll 1$.

For $0 < c < 1$, we are able to construct a corresponding singular limit ($\varepsilon \to 0$) travelling wave profile with a shock (i.e., a fast jump) as a concatenation of segments of the two limiting systems, the layer and the reduced problem; see Fig. 2.15. Again, Fenichel theory (see Chap. 3) provides the results on the existence of a nearby travelling wave for sufficiently small perturbations $0 < \varepsilon \ll 1$. Note that for $0 < c < 1$, the reduced flow is away from the contact point F (4.29) while for $c > 1$ it is towards F.

Remark 4.8 *The case $c = 1$ where the regular jump point condition (4.31) is violated will be discussed in Chap. 6.*

Remark 4.9 *Self-similar solutions in a Dafermos regularisation of a hyperbolic PDE can be analysed in a similar fashion. For details, we refer the reader to, e.g., Schecter and Szmolyan [97].*

4.4 Centre Manifold Reduction Near Contact Points

In a neighbourhood of a contact point $z \in F$, there exists a local $(k + 1)$-dimensional centre manifold W^c. The *Centre Manifold Theorem* (see, e.g., [19, 63]) states that the flow on this centre manifold W^c captures all the essential (nonlinear) dynamics near such a contact point. The following result gives such a local centre manifold reduction of system (3.10).

Theorem 4.1 *Given a singularly perturbed system in general form* (3.10) *with a set of contact points $F \subset S$ as by Definition 4.1. Assume local co-ordinates $z = (x, y)^\top$, $x \in \mathbb{R}^k$ and $y \in \mathbb{R}^{n-k}$ near a contact point $z \in F$ such that $D_y f(x, y)$ is a regular $(n - k)$-dimensional square matrix, i.e., the k-dimensional critical manifold S is locally a graph over the x-coordinate chart. Let $l, r \in \mathbb{R}^{n-k}$ denote the left and right null vector of the matrix $DfN|_{z \in F}$, respectively, and let P denote a corresponding $(n - k) \times (n - k - 1)$ matrix such that the set of column vectors $\{r, P\}$ forms a basis of $\mathbb{R}^{n-k}$ at $z \in F$ with $l \cdot P = \mathbb{O}_{n-k-1}$.*

There exists a $(k+1)$-dimensional centre manifold W^c near such a contact point $z \in F$. The flow on this centre manifold W^c is locally equivalent to the $(k + 1)$-dimensional system

$$\begin{pmatrix} x' \\ u' \end{pmatrix} = \tilde{N}\tilde{f} + \varepsilon\tilde{G} = \begin{pmatrix} N^x r + N^x P W_0(x, u) \\ lDfNr + lDfNPW_0(x, u) \end{pmatrix} u + \varepsilon \begin{pmatrix} \tilde{G}^x \\ \tilde{G}^u \end{pmatrix}, \quad (4.32)$$

where $u \in \mathbb{R}$ is a local coordinate in the direction of the contact spanned by a nontrivial column vector of the rank one matrix $N\,adj(Df\,N)|_F$, i.e., by Nr.

Proof First, we straighten locally the critical manifold S as in the proof of Lemma 3.3: the coordinate transformation $v = f(x, y)$ with $y = K(x, v)$ as its inverse, transforms system (3.10) locally to

$$\begin{pmatrix} x' \\ v' \end{pmatrix} = \begin{pmatrix} N^x(x, K(x, v))) \\ DfN(x, K(x, v)) \end{pmatrix} v + \varepsilon \begin{pmatrix} G^x(x, K(x, v))) \\ DfG(x, K(x, v)) \end{pmatrix}, \quad (4.33)$$

where the critical manifold is now given by the set $S = \{v = 0\}$.

Next, we isolate the $(k+1)$-dimensional centre subspace at a contact point $z \in F$ as follows[5]: without loss of generality, we assume that such a contact point of system (4.33) is located at the origin.[6] We introduce the linear co-ordinate transformation

$$\begin{pmatrix} u \\ w \end{pmatrix} = \begin{pmatrix} l \\ Q \end{pmatrix} v, \quad (4.34)$$

with corresponding inverse

$$v = \begin{pmatrix} r & P \end{pmatrix} \begin{pmatrix} u \\ w \end{pmatrix} = ru + Pw, \quad (4.35)$$

where $u \in \mathbb{R}$, $w \in \mathbb{R}^{n-k-1}$, l and r denote left and right null vectors of the matrix DfN, Q is a corresponding $(n - k - 1) \times (n - k)$ matrix and P is a corresponding $(n - k) \times (n - k - 1)$ matrix such that

$$\begin{pmatrix} r & P \end{pmatrix} \begin{pmatrix} l \\ Q \end{pmatrix} = \begin{pmatrix} l \\ Q \end{pmatrix} \begin{pmatrix} r & P \end{pmatrix} = \mathbb{I}_{n-k}, \quad (4.36)$$

---

[5] This part follows the same steps as in the proof of Lemma 4.1.

[6] A simple translation of the contact point suffices.

which gives the new system

$$\begin{pmatrix} x' \\ u' \\ w' \end{pmatrix} = \begin{pmatrix} N^x \\ lDfN \\ QDfN \end{pmatrix} (r \ P) \begin{pmatrix} u \\ w \end{pmatrix} + \varepsilon \begin{pmatrix} G^x \\ lDfG \\ QDfG \end{pmatrix}. \qquad (4.37)$$

The corresponding Jacobian evaluated at the contact point (located at the origin) is given by

$$\begin{pmatrix} \mathbb{O}_{k,k} & N^x r & N^x P \\ \mathbb{O}_{1,k} & 0 & \mathbb{O}_{1,n-k-1} \\ \mathbb{O}_{n-k-1,k} & \mathbb{O}_{n-k-1,1} & Q\,DfN\,P \end{pmatrix}. \qquad (4.38)$$

The $(k+1)$-dimensional generalised nullspace E^c of (4.38) is spanned by the (x,u)-coordinate chart and the contact happens in the direction of the vector $(x,u) = (N^x r, 0)$. The $(n-k-1)$-dimensional square matrix $QDfNP$ encodes the $(n-k-1)$ nonzero eigenvalues, i.e., the matrix $QDfNP$ has full rank. Thus, by the implicit function theorem, the corresponding centre manifold W^c is locally a graph over the (x,u)-coordinate chart, i.e.,

$$w = \tilde{W}(x,u,\varepsilon) = W_0(x,u) + \varepsilon W_1(x,u,\varepsilon) \qquad (4.39)$$

with $W_0(x,u) = O(\|(x+u)\|)$. In fact, $W_0 = u\tilde{W}_0(x,u)$ due to the flatness of S, i.e., $\tilde{W}(x,0,0) = \mathbb{O}_{n-k-1,1}$. The flow on this centre manifold W^c is described by system (4.32). $\qquad \square$

Remark 4.10 *In general, the nonlinear coordinate transformation $v = f(x,y)$ that straightens the critical manifold S is not necessary to obtain such a local centre manifold reduction. A corresponding linear coordinate transformation $v = Df z$ with constant coefficient matrix $Df|_{z\in F}$ such that $\{v = 0\}$ corresponds to the tangent space $T_z S$ at the contact point $z \in F$ (assumed without loss of generality to be located at the origin) would suffice. The corresponding flow on W^c is then locally equivalent to the $(k+1)$-dimensional system*

$$\begin{pmatrix} x' \\ u' \end{pmatrix} = \tilde{N}(x,u)\tilde{f}(x,u) + \varepsilon\tilde{G}(x,u), \qquad (4.40)$$

where $u \in \mathbb{R}$ is the same local coordinate defined by (4.34). The main difference between system (4.32) and (4.40) is that the function $\tilde{f}(x,u)$ is, in general, nonlinear for system (4.40).

4.5 Regular Jump Point Results for $0 < \varepsilon \ll 1$

As stated in Theorem 4.1, the local flow near a contact point is captured by system (4.32) on the corresponding $(k+1)$-dimensional centre manifold W^c. Local results on regular jump points are best stated for a local canonical form

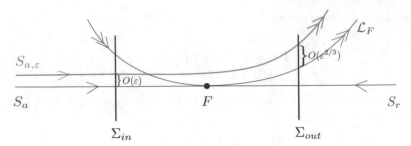

Fig. 4.1 Sketch of the local flow near a regular jump point F where the critical manifold S and the fast fibre $\mathcal{L}_F$ have contact of order one and the reduced flow is towards F. The corresponding attracting slow manifold $S_{a,\varepsilon}$ which is $O(\varepsilon)$ close to S_a in the cross section Σ_{in}, extended by the flow past the contact point F, exits $O(\varepsilon^{2/3})$ close to the fast fibre $\mathcal{L}_F$ in the cross section Σ_{out}

near such an order one contact point F obtained after some additional local coordinate transformations on system (4.32)[7]:

Lemma 4.3 *Given the nonstandard singularly perturbed system* (3.10) *under the assumption that the critical manifold S has order one contact with the layer flow along F and that the reduced flow is towards F. Then there exists a $(k+1)$-dimensional centre manifold $\tilde{W}_c$ in a neighbourhood of a regular jump point $z \in F$ and a corresponding local $(k+1)$-dimensional coordinate system in which* (3.10) *restricted to $\tilde{W}_c$ takes the local canonical form*

$$\begin{pmatrix} \tilde{x}_1' \\ \tilde{\xi}' \\ \tilde{u}' \end{pmatrix} = \tilde{N}\tilde{f} + \varepsilon\tilde{G} = \begin{pmatrix} 1 + O(\tilde{x}_1, \tilde{u}) \\ O(\tilde{x}_1, \tilde{u}) \\ \tilde{x}_1 + O(\tilde{u}) \end{pmatrix} u + \varepsilon \begin{pmatrix} \tilde{G}^{x_1}(\tilde{x}_1, \tilde{\xi}, \tilde{u}, \varepsilon) \\ \tilde{G}^{\xi}(\tilde{x}_1, \tilde{\xi}, \tilde{u}, \varepsilon) \\ 1 + O(\tilde{x}_1, \tilde{u}, \varepsilon) \end{pmatrix}, \quad (4.41)$$

with local coordinates $\tilde{x}_1 \in \mathbb{R}$, $\tilde{\xi} \in \mathbb{R}^{k-1}$, $\tilde{u} \in \mathbb{R}$ and arbitrary functions $\tilde{G}^{x_1}$ and $\tilde{G}^{\xi}$.

For small positive $\rho > 0$, we define local cross sections $\Sigma_{in} = \{(\tilde{x}_1, \tilde{\xi}, \tilde{u}) \in \mathbb{R}^{k+1} : \tilde{x}_1 = -\rho\}$ and $\Sigma_{out} = \{(\tilde{x}_1, \tilde{\xi}, \tilde{u}) \in \mathbb{R}^{k+1} : \tilde{x}_1 = \rho\}$ in a suitable compact neighbourhood $\mathcal{B}_S = \{(\tilde{x}_1, \tilde{\xi}, \tilde{u}) \in \mathbb{R}^{k+1} : (\tilde{\xi}, \tilde{u}) \in I \subset \mathbb{R}^k\}$ of the critical manifold S; see Fig. 4.1. The following result describes the local flow near regular jump points explaining the transition from slow to fast motion.

Theorem 4.2 *For system* (4.41) *there exists an $\varepsilon_0 > 0$ such that for $\varepsilon \in (0, \varepsilon_0)$ the following holds:*

1. *The manifold $S_{a,\varepsilon}$ intersects Σ_{out} along a graph $\tilde{u} = h_{out}(\tilde{\xi}, \varepsilon) = h_{out}^0(\tilde{\xi}) + O(\varepsilon^{2/3})$, where $h_{out}^0(\tilde{\xi})$ corresponds to the intersection of the subset of the*

---

[7] This is left as an exercise for the reader; see, e.g., [103].

fast layer fibration $\mathcal{L}_F \subset \mathcal{L}$ that has contact of order one along F with the section Σ_{out}.

2. *The (local) section Σ_{in} is mapped to an exponentially thin strip around $S_{a,\varepsilon} \cap \Sigma_{out}$, i.e., its with in $\tilde{u}$ direction is $O(e^{-k/\varepsilon})$ where $k > 0$ is a positive constant.*

3. *The map $\Pi : \Sigma_{in} \to \Sigma_{out}$ has the form*

$$\Pi \begin{pmatrix} \tilde{\xi} \\ \tilde{u} \end{pmatrix} = \begin{pmatrix} \tilde{G}(\tilde{\xi}, \tilde{u}, \varepsilon) \\ h_{out}(\tilde{G}(\tilde{\xi}, \tilde{u}, \varepsilon), \varepsilon) + O(e^{-k/\varepsilon}) \end{pmatrix} \tag{4.42}$$

with $\tilde{G}(\tilde{\xi}, \tilde{u}, \varepsilon) = \tilde{G}_0(\tilde{\xi}) + O(\varepsilon \ln \varepsilon)$ where $\tilde{G}_0(\tilde{\xi}) = \tilde{\xi} + O(\rho^3)$ is induced by the reduced flow on S_a from $\Sigma_{in} \cap S$ to F and $h_{out}(\tilde{G}(\tilde{\xi}, \tilde{u}, \varepsilon), \varepsilon) = h_{out}^0(\tilde{\xi}) + O(\varepsilon^{2/3})$.

Remark 4.11 *Due to the possible local transformation of a general singularly perturbed system with contact manifold F to the standard form (1.1) with folded critical manifold S, all local results on regular jump points in the standard case apply to the general case. For the corresponding standard results we refer to, e.g., [57, 78, 103, 111].*

To prove these results one appeals to the 'blow-up' technique for nilpotent singularities [16, 27, 61]. We will discuss the 'blow-up' technique briefly for the autocatalator model in Sect. 5.3.

The important insight of Theorem 4.2 is that the $(k+1)$-dimensional centre flow of system (4.41) restricted to an ε dependent neighbourhood $\mathcal{B}_F$ of the $(k-1)$-dimensional manifold F of order one contact points, i.e., for

$$\mathcal{B}_F = \{(\tilde{x}_1, \tilde{\xi}, \tilde{u}) \in \mathbb{R}^{k+1} : |\tilde{x}_1| < k_1 \varepsilon^{1/3}, \, |\tilde{u}| < k_2 \varepsilon^{2/3}, \, \tilde{\xi} \in \tilde{I} \subset \mathbb{R}^{k-1}\}$$

with $k_1, k_2 > 1$, is to leading order captured by the two-dimensional system

$$\begin{aligned} \tilde{x}_1' &= \tilde{u} \\ \tilde{u}' &= \tilde{x}_1 \tilde{u} + 1 \,. \end{aligned} \tag{4.43}$$

The other slow variables $\tilde{\xi} \in \mathbb{R}^{k-1}$ act as 'parameters' (to leading order), i.e., $\tilde{\xi}' = 0$ in this ε-dependent neighbourhood. Introducing a new coordinate $\tilde{y}_1 = \tilde{u} - \tilde{x}_1^2/2$ transforms this two-dimensional system (4.43) to

$$\begin{aligned} \tilde{x}_1' &= \tilde{y}_1 + \frac{\tilde{x}_1^2}{2} \\ \tilde{y}_1' &= 1 \,, \end{aligned} \tag{4.44}$$

which is equivalent to the well-known Riccati equation $\tilde{x}_1' = t + \tilde{x}_1^2/2$ and describes the local flow near a regular jump point.

Chapter 5
Relaxation Oscillations in the General Setting

As mentioned in the previous chapter, the local dynamics near regular jump points indicate one possibility for solutions of a general singular perturbation problem (3.10), respectively, (3.15) to switch from slow to fast dynamics (or vice versa) which is key for any global relaxation oscillatory behaviour.

Definition 5.1 *A singular relaxation cycle Γ is a concatenation of orbit segments of the reduced problem and the layer problem that form a closed loop.*

There is a multitude of possibilities to construct such (global) singular orbits, especially in the general setting which provides additional possibilities for global return mechanisms (compared to the standard setting). For illustrative purposes, we focus here on one specific class of singular periodic relaxation orbits, where $k = n - 1$, i.e., the layer problem is only one-dimensional, and a transition from slow to fast dynamics happens at a regular jump point while the transition from fast to slow dynamics happens in a normally hyperbolic regime of the critical manifold S. This setting is sufficient to create two-stroke relaxation oscillators as introduced in Chap. 2.

Proposition 5.1 *Given a general singularly perturbed system (3.10), respectively, (3.15) with a $(n-1)$-dimensional critical manifold $S = S_a \cup F \cup S_r$ where F is a $(n-2)$-dimensional manifold of order one contact points. Let $\Gamma = \gamma_s \cup \gamma_f$ denote a singular relaxation cycle that is a concatenation of an orbit segment γ_s of the corresponding reduced problem from a point $p_l \in S_a$ to a regular jump point $p_j \in F$ and an orbit segment γ_f of the corresponding layer problem that connects from $p_j \in F$ back to $p_l \in S_a$. Assume further that the projection of a sufficiently small open segment of the set of contact points F through p_j along fast fibres onto S_a is transverse to the reduced flow near p_l (see Fig. 5.1).*

For sufficiently small $\varepsilon \ll 1$, there follows the existence of a unique attracting relaxation cycle Γ_ε that is $O(\varepsilon^{2/3})$-close to the singular relaxation cycle Γ (in the sense of a Hausdorff distance).

© Springer Nature Switzerland AG 2020

M. Wechselberger, *Geometric Singular Perturbation Theory Beyond the Standard Form*, Frontiers in Applied Dynamical Systems: Reviews and Tutorials 4, https://doi.org/10.1007/978-3-030-36399-4_5

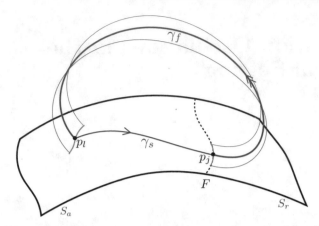

Fig. 5.1 Sketch of a singular relaxation cycle $\Gamma = \gamma_s \cup \gamma_f$ that is a concatenation of an orbit segment γ_s of the corresponding reduced problem from a point $p_l \in S_a$ to a regular jump point $p_j \in F$ and an orbit segment γ_f of the corresponding layer problem that connects from $p_j \in F$ back to $p_l \in S_a$. The green segment through p_l represents the projection along fast fibres of the green segment of contact points (regular jump points) through p_j that connect back to S_a via fast fibres

Proof For the general case of two-stroke oscillators for $n = 2$ see [47] and [45]. Proofs on relaxation oscillations in the standard case are presented in [57] for $n = 2$ and [103] for $n = 3$ (which also holds for $n \geq 3$). □

Remark 5.1 *In the general planar case $n = 2$, we would like to point to the existence results on slow-fast cycles formed by an arbitrary sequence of slow and fast orbit segments derived by de Maesschalck, Dumortier and Roussarie [23] using the notion of a* slow divergence integral.

Remark 5.2 *In the standard planar case $n = 2$, relaxation oscillations cannot be found if only one regular jump point is involved. Thus two-stroke oscillators for $n = 2$ form a genuine example of relaxation oscillations.*

5.1 Two-Stroke Oscillator Revisited

Recall the 'stick-slip' (two-stroke) oscillator model (2.28) that can be written as a singularly perturbed system (2.36) which is of the general form (3.10) with

$$z = \begin{pmatrix} x \\ y \end{pmatrix} \in \mathbb{R}^2, \ N(x,y) = \begin{pmatrix} y \\ -x+y \end{pmatrix}, \ f(x,y) = 1-y \qquad (5.1)$$

and

$$G(x, y, \varepsilon) = \begin{pmatrix} 0 \\ -1 \end{pmatrix}. \tag{5.2}$$

Layer Problem:
Here, the layer problem is the following system:

$$\begin{pmatrix} x' \\ y' \end{pmatrix} = N(x, y) f(x, y) = \begin{pmatrix} y \\ -x + y \end{pmatrix} (1 - y). \tag{5.3}$$

The set of singularities of this layer problem is given by

$$S_0 = \{y = 1\} \cup \{(0, 0)^\top\}. $$

Recall from Sect. 2.2.2 that the isolated singularity at the origin defined by $N(x, y) = 0$ is an unstable focus. The subset

$$S = \{(x, y) \in \mathbb{R}^2 : y = 1\} \subset S_0 \tag{5.4}$$

forms the one-dimensional critical manifold of this problem which is a regular level set. The Jacobian of the layer problem evaluated along $(x, y) \in S$ is given by

$$N Df|_S = \begin{pmatrix} 0 & -1 \\ 0 & x - 1, \end{pmatrix}$$

which has one trivial eigenvalue and $\lambda_1 = Df N|_S = x - 1$, $\forall x \in \mathbb{R}$, as its nontrivial eigenvalue. Thus S loses normal hyperbolicity at a single contact point

$$F = \{(x, y) \in S : x = 1\} = \{(1, 1)^\top\} \tag{5.5}$$

and $S = S_a \cup F \cup S_r$ consists of an attracting branch S_a for $x < 1$ and a repelling branch for $x > 1$. The conditions (4.6) of Lemma 4.1 are fulfilled, i.e.,

$$\operatorname{rk} Df|_F = n - k = 1,$$

$$l \cdot (D^2 f(Nr, Nr) + Df DN(Nr, r)) = Df DN N|_F = (0 \ -1) \begin{pmatrix} 0 & 1 \\ -1 & 1 \end{pmatrix} \begin{pmatrix} 1 \\ 0 \end{pmatrix} = 1 \neq 0,$$

which shows that the contact is of order one. Note that the nontrivial eigenvalue $\lambda_1(x) = \det(Df N)|_S = x - 1$ is parameterised by x along S, and we have

$$\lambda_1'(x = 1) = 1 \neq 0, \tag{5.6}$$

i.e., the sign of $\det(Df N)|_S$ changes as we cross F transversally as indicated by Lemma 4.2. Note further that the fast rotations around the isolated singularity $(0, 0)^\top$ cause the order one contact of the layer flow with S at F.

Remark 5.3 *The layer problem is for $y < 1$ equivalent to the linear system*

$$\begin{pmatrix} x' \\ y' \end{pmatrix} = N(x,y) = \begin{pmatrix} 0 & 1 \\ -1 & 1 \end{pmatrix} \begin{pmatrix} x \\ y \end{pmatrix}, \tag{5.7}$$

i.e., we can divide out the common factor $f(x,y) = 1 - y$.

Reduced Problem:

Here, we study the corresponding desingularised problem (4.20) which is given by

$$\begin{pmatrix} \dot{x} \\ \dot{y} \end{pmatrix} = \Pi_d^S G(x,y,0) = \left(-(x-1) \begin{pmatrix} 1 & 0 \\ 0 & 1 \end{pmatrix} + \begin{pmatrix} 0 & -y \\ 0 & x-y \end{pmatrix} \right) \begin{pmatrix} 0 \\ -1 \end{pmatrix}$$

evaluated along the critical manifold S defined in (5.4). Since S is a graph over the x-coordinate chart, we study the desingularised flow in this single coordinate chart, i.e.,

$$\dot{x} = 1; \tag{5.8}$$

compare with the corresponding reduced equation (2.41). Since $\lambda_1 = x-1$, the flow of this desingularised system has to be reversed on S_r, i.e., for $x > 1$, to obtain the corresponding reduced flow. Note further, that the single nontrivial condition (4.22) for a regular jump point, i.e.,

$$N \operatorname{adj} (Df\, N) Df\, G(x,y,0)|_{(1,1)} = \begin{pmatrix} 1 \\ 0 \end{pmatrix} \neq \mathbb{O}_2 \tag{5.9}$$

is satisfied. Hence, the reduced flow is towards the contact point $F = (1,1)^\top$ and we expect a finite time blow-up near this contact point which allows a switch from the slow to the fast dynamics in the full system.

Relaxation Oscillator:

From the above analysis, we conclude the existence of a singular relaxation cycle Γ which is a concatenation of an orbit segment of the reduced problem connecting from a point $p_l = (x_l < 1, 1)$ on S_a to the jump point $p_j = F = \{(1,1)^\top\}$ and a fast fibre segment arc connecting from $p_j = F$ back to $p_l \in S_a$; see Fig. 5.2.[1]

By Proposition 5.1, this singular relaxation cycle perturbs to a nearby attracting relaxation cycle Γ_ε as shown in Fig. 2.6 (lower panel), and the corresponding time traces are shown in Fig. 2.6 (upper panel).

5.2 Three Component Negative Feedback Oscillator Revisited

Recall the negative feedback loop model (2.42) which can be written as a singularly perturbed system in general form (3.10), respectively, (3.15) with

---

[1] Although the layer problem is linear, we can only define the point p_l implicitly.

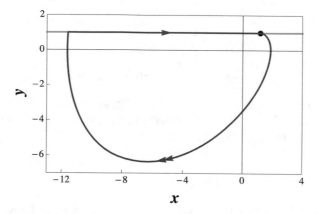

Fig. 5.2 Singular relaxation cycle of system (2.36) as a concatenation of a slow (blue) and a fast (red) segment (singular 'two-stroke' oscillator). The transition from slow to fast motion happens near the contact point F (black dot) which is a regular jump point

$$N(x,y,z) = \begin{pmatrix} \alpha_1(\dfrac{1}{1+z^2} - x) \\ \alpha_2 x - 1 \\ \alpha_3(y-z) \end{pmatrix}, \ f(x,y,z) = y \qquad (5.10)$$

and

$$G(x,y,z,\varepsilon) = \begin{pmatrix} \alpha_1(\dfrac{1}{1+z^2} - x) \\ \alpha_2 x \\ \alpha_3(y-z) \end{pmatrix}. \qquad (5.11)$$

Layer Problem:

$$\begin{pmatrix} x' \\ y' \\ z' \end{pmatrix} = N(x,y,z)f(x,y,z) = \begin{pmatrix} \alpha_1(\dfrac{1}{1+z^2} - x) \\ \alpha_2 x - 1 \\ \alpha_3(y-z) \end{pmatrix} y. \qquad (5.12)$$

The set of singularities of the layer problem is given by

$$S_0 = \{y = 0\} \cup \{(\alpha_2^{-1}, \sqrt{\alpha_2 - 1}, \sqrt{\alpha_2 - 1})^\top\}.$$

This set S_0 contains a subset

$$S = \{y = 0\} \subset S_0, \qquad (5.13)$$

which is a regular level set and it forms the two-dimensional critical manifold of this problem; it is a graph over the single coordinate chart $(x, z) \in \mathbb{R}_0^+ \times \mathbb{R}_0^+$. The Jacobian of the layer problem evaluated along $(x, y, z) \in S$ is given by

$$
NDf|_S = \begin{pmatrix} 0 & \alpha_1(\dfrac{1}{1+z^2} - x) & 0 \\ 0 & \alpha_2 x - 1 & 0 \\ 0 & -\alpha_3 z & 0, \end{pmatrix}
$$

which has two trivial eigenvalues and $\lambda_1 = DfN|_S = \alpha_2 x - 1$ as its nontrivial eigenvalue. Thus S loses normal hyperbolicity for $x = \alpha_2^{-1}$, and $S = S_a \cup \tilde{F} \cup S_r$ consists of an attracting branch S_a for $x < \alpha_2^{-1}$, a repelling branch for $x > \alpha_2^{-1}$ and a one-dimensional submanifold

$$
\tilde{F} = \{(x, y, z) \in S \; : \; x = \alpha_2^{-1}\} \tag{5.14}
$$

of contact points. For this problem, the conditions (4.6) in Lemma 4.1 give

$$
\mathrm{rk}\, Df|_{\tilde{F}} = n - k = 1,
$$

$$
Df DNN|_{\tilde{F}} = (0\ 1\ 0) \begin{pmatrix} -\alpha_1 & 0 & -2\alpha_1 \dfrac{z}{(1+z^2)^2} \\ \alpha_2 & 0 & 0 \\ 0 & \alpha_3 & -\alpha_3 \end{pmatrix} \begin{pmatrix} \dfrac{\alpha_1(\alpha_2 - (1+z^2))}{\alpha_2(1+z^2)} \\ 0 \\ -\alpha_3 z \end{pmatrix}
$$

$$
= \frac{\alpha_1(\alpha_2 - (1+z^2))}{1+z^2},
$$

which shows that all contact points $z \in F = \tilde{F} \backslash C$ are of order one, where

$$
C = \{(x, y, z) \in \tilde{F} \; : \; z = \sqrt{\alpha_2 - 1}\} = \{(\alpha_2^{-1}, 0, \sqrt{\alpha_2 - 1})^\top\}. \tag{5.15}
$$

Note that the nontrivial eigenvalue $\lambda_1(x, z) = \det(DfN)|_S = \alpha_2 x - 1$ is independent of z. Furthermore, any (one-dimensional) path $c(s) = (\alpha_2^{-1} + s, 0, z_0)$, $s \in [-s_0, s_0]$, along S crosses $\tilde{F}$ transversally and we have

$$
\frac{d}{ds}\lambda_1(c(s))|_{s=0} = \alpha_2 > 0.
$$

Thus the sign of $\det(DfN)|_S$ changes as we cross this set $\tilde{F}$ transversally which confirms the splitting of the critical manifold $S = S_a \cup \tilde{F} \cup S_r$.

Remark 5.4 *The isolated contact point $C \in \tilde{F}$ (5.15) is of order two; see Fig. 5.3 for a visual confirmation.[2] This contact of order two at the point C is equivalent to a local codimension-two 'cusp' structure in a standard singular perturbation problem; see [12] for details on such a locally equivalent standard problem.*

Another local codimension-two structure in a standard singular perturbation problem, the slow-fast Bogdanov-Takens point, has been studied in [21]; see also [70] for a general exposition of 'bifurcation without parameters'.

---

[2] Otherwise, one has to show that the third derivative does not vanish; see Definition 4.2.

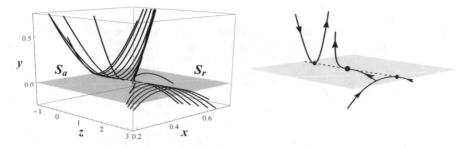

Fig. 5.3 Layer flow, (2.43) for $\varepsilon = 0$, near the line of regular jump points $F : x = \alpha_2^{-1}$, $\alpha_2 > 1$. As can be seen, the line F consists of order one contact points with the exception of the contact point at $z = \sqrt{\alpha_2 - 1}$ (indicated with a red dot in right panel) which has contact order two

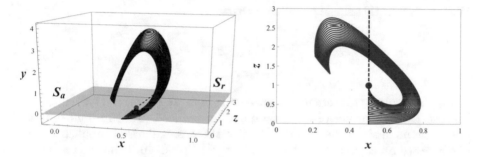

Fig. 5.4 Global return mechanisms of the layer flow, (2.43) for $\varepsilon = 0$, starting from a segment of the line of regular jump points $F : x = \alpha_2^{-1}$ (for $0 \le z \le \sqrt{\alpha_2 - 1}$) where the reduced flow is direct towards F, the layer flow returns back to S_a. (Left) 3D phase space representation; (Right) corresponding projection onto (x, z)-space

Finally, we look at the isolated singularity $(\alpha_2^{-1}, \sqrt{\alpha_2 - 1}, \sqrt{\alpha_2 - 1})^{\top}$ of the layer problem defined by $N(x, y, z) = 0$. It is of saddle-focus type with one negative real eigenvalue and a pair of complex conjugate eigenvalues with positive real part.[3] Since it is a hyperbolic equilibrium, it will persist as a nearby saddle-focus in the full system; see Remark 3.11. It is this isolated singularity that creates a global return mechanisms for the regular jump solutions along F back to the attracting manifold S_a as shown in Fig. 5.4.[4]

---

[3] Calculate the characteristic polynomial for this singularity: based on the assumptions on the parameters $(\alpha_1, \alpha_2, \alpha_3)$, *Descartes Sign Rule* shows that there are no real positive roots and the *Routh–Hurwitz Test* shows that not all roots have negative real parts.

[4] Showing this layer problem connections analytically is a nontrivial task.

Reduced Problem:
We study the corresponding desingularised problem (4.20) evaluated along
the critical manifold S (5.13) which is given by

$$
\begin{pmatrix} \dot{x} \\ \dot{y} \\ \dot{z} \end{pmatrix} = \Pi_d^S G(x, y, z, 0)|_S
$$

$$
= \left((1 - \alpha_2 x) \begin{pmatrix} 1 & 0 & 0 \\ 0 & 1 & 0 \\ 0 & 0 & 1 \end{pmatrix} + \begin{pmatrix} 0 & \alpha_1(\dfrac{1}{1+z^2} - x) & 0 \\ 0 & \alpha_2 x - 1 & 0 \\ 0 & -\alpha_3 z & 0 \end{pmatrix} \right) \begin{pmatrix} \alpha_1(\dfrac{1}{1+z^2} - x) \\ \alpha_2 x \\ -\alpha_3 z \end{pmatrix}.
$$

Since S is a graph over the (x, z)-coordinate chart, we study the desingularised flow in this single coordinate chart:

$$
\begin{aligned}
\dot{x} &= \alpha_1\left(\frac{1}{1+z^2} - x\right) \\
\dot{z} &= -\alpha_3 z.
\end{aligned}
\tag{5.16}
$$

This desingularised system has an equilibrium at $(x, z) = (1, 0) \in S_r$. It
is also an equilibrium of the corresponding reduced problem, and it is an
unstable node since the flow direction of the desingularised problem has to
be reversed on S_r to obtain that of the reduced problem. It has no particular
influence on the relaxation oscillations observed and it is independent of the
system parameters α_1, α_2 and α_3.

As shown in Fig. 2.11, the reduced flow in the neighbourhood of $\tilde{F}$ is
towards $\tilde{F}$ for $z < \sqrt{\alpha_2 - 1}$, away from $\tilde{F}$ for $z < \sqrt{\alpha_2 - 1}$ and there is a
tangency with $\tilde{F}$ at the point C given by $z = \sqrt{\alpha_2 - 1}$. This indicates that
each order one contact point $z \in F$ is a regular jump point which is confirmed
by condition (4.22), i.e.,

$$
N \mathrm{adj}\,(Df\,N) Df\,G(x, y, 0)|_{\tilde{F}} = \begin{pmatrix} \alpha_1 \dfrac{\alpha_2 - (1+z^2)}{\alpha_2(1+z^2)} \\ 0 \\ -\alpha_3 z \end{pmatrix} \neq \mathbb{O}_3, \quad \forall z \geq 0. \tag{5.17}
$$

Due to the change of orientation of the reduced flow along $\tilde{F}$, the tangency of
the reduced flow at the *cusp*-like singularity $C \in \tilde{F}$ is necessary by continuity;
see Remark 5.4.

Relaxation Oscillator:
We are able to construct a singular relaxation cycle Γ which is a concatenation
of an orbit segment of the reduced problem connecting from a point $p_l =$
$(x_l, 0, z_l)$ on S_a to a regular jump point $p_j = (\alpha_2^{-1}, 0, z_j < \sqrt{\alpha_2 - 1}) \in F$
where the reduced flow is towards F and a fast fibre segment arc connecting

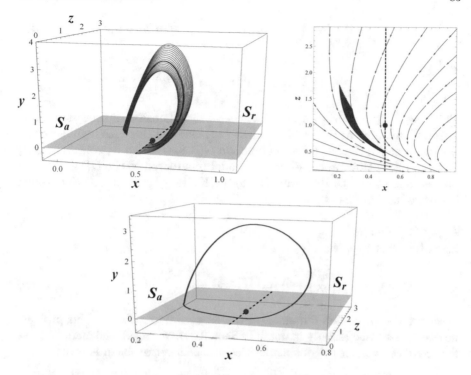

Fig. 5.5 Singular limit flows of the three component negative feedback oscillator model with $\alpha_1 = 0.2$, $\alpha_2 = 2$, $\alpha_3 = 0.2$ projected onto (x, z) plane: (Upper Left) layer flow, (2.43) for $\varepsilon = 0$, from line segment of the regular jump points $F : x = \alpha_2^{-1}$ for $z < 1$ until the return on S_a. (Upper Right) desingularised flow (5.16), from the 'landing points' of the layer problem on S_a to the line of regular jump points F. These two panels show that a 'return map' can be defined and it is clearly a contraction, i.e., indicates the existence of a unique attracting singular relaxation oscillator Γ shown in the (Lower) panel

from $p_j = (\alpha_2^{-1}, 0, z_j < \sqrt{\alpha_2 - 1}) \in F$ back to $p_l = (x_l, 0, z_l) \in S_a$; see Fig. 5.5.

By Proposition 5.1, this singular relaxation cycle Γ perturbs to a nearby attracting relaxation cycle Γ_ε. The proof based on the *contraction mapping principle* for the associated return map near Γ is omitted, but Fig. 5.5 provides a compelling visual 'proof' (note the strong contraction of the projection of the contact point segment towards F shown in the upper right panel).

5.3 Autocatalator Revisited

Recall[5] the autocatalator model (2.54) that can be written as a singularly perturbed system in the general form (3.10) with

---

[5] A detailed analysis can be found in [36].

$$z = \begin{pmatrix} a \\ b \end{pmatrix} \in \mathbb{R}^2, \ N(a,b) = \begin{pmatrix} -1 \\ 1 \end{pmatrix}, \ f(a,b) = ab^2 \qquad (5.18)$$

and

$$G(a,b,\varepsilon) = \begin{pmatrix} 0 \\ -b \end{pmatrix} + \varepsilon \begin{pmatrix} \mu - a \\ a \end{pmatrix}. \qquad (5.19)$$

This singularly perturbed system in general form consists of processes evolving on three different time scales—ultra-fast, fast and slow—as highlighted by the two terms in the perturbation function $G(a,b,\varepsilon)$ which represent the $O(\varepsilon)$ fast and $O(\varepsilon^2)$ slow processes in this model (compared to the $O(1)$ ultra-fast processes modelled by the term $N(a,b)f(a,b)$. This is the first distinguishing feature of this oscillator model.

Layer Problem:
Here, the layer problem

$$\begin{pmatrix} a' \\ b' \end{pmatrix} = N(a,b)f(a,b) = \begin{pmatrix} -1 \\ 1 \end{pmatrix} ab^2 \qquad (5.20)$$

describes the super-fast dynamics of the autocatalator model. This problem has been discussed already in detail in Sect. 2.2.4. We recall the main findings for convenience. The set of singularities of this layer problem is given by

$$S = S_a \cup S_d = \{a = 0\} \cup \{b = 0\},$$

i.e., the set S is the union of two one-dimensional critical manifolds S_a and S_d that intersect at the origin. This is another distinguishing feature of this model. While this does not fit Assumption 3.2, i.e., it violates the existence of a connected critical manifold S that is the regular zero level set of the function f, it does not cause a problem for the application of GSPT as previously indicated in Remark 3.3.[6]

The Jacobian of the layer problem evaluated along $(a,b) \in S_a$ is given by

$$NDf|_{S_a} = \begin{pmatrix} -b^2 & 0 \\ b^2 & 0 \end{pmatrix},$$

which has one trivial eigenvalue and $\lambda_1 = DfN|_{S_a} = -b^2$, $\forall b \in \mathbb{R}$, as its nontrivial eigenvalue. Thus S_a loses normal hyperbolicity for $b = 0$, and it consists of two attracting branches: one for $b > 0$ and another for $b < 0$. From a modelling point of view, we will ignore the negative (i.e., biochemical irrelevant) branch. Similarly, the Jacobian of the layer problem evaluated along $(a,b) \in S_d$ is given by

$$NDf|_{S_d} = \begin{pmatrix} 0 & 0 \\ 0 & 0 \end{pmatrix},$$

---

[6] The intersection point at the origin can also be analysed with GSPT techniques.

which has one trivial eigenvalue and $\lambda_1 = Df N|_{S_d} = 0$, $\forall a \in \mathbb{R}$, as its nontrivial eigenvalue. Thus S_d is a line of non-hyperbolic equilibria which is another distinguishing feature of this model, since loss of normal hyperbolicity usually occurs along codimension-one submanifold(s) of a critical manifold.

Remark 5.5 *Away from the origin, the set S_a is a regular level set, i.e., Df has full rank along S_a, as assumed in our analysis. On the other hand, S_d consists entirely of critical points, i.e., Df vanishes along S_d. We will explain shortly how one deals with such a line of nilpotent equilibria.*

The layer flow of (5.20) is particularly simple here since the constant vector $N(a, b) = (-1, 1)^\top$ represents the vector field of the layer problem away from S. This leads to straight ultra-fast fibres connecting the two branches S_a and S_d as shown in Fig. 2.12 (right), i.e., we have $db/da = -1$ and the attracting manifold S_a is approached along these lines.

Reduced Problem Along S_a:
The 'fast' motion along the attracting branch S_a (compared to the super-fast motion towards S_a described by the layer problem) is studied in the corresponding desingularised problem (4.20) which is given by

$$\begin{pmatrix} \dot{a} \\ \dot{b} \end{pmatrix} = \Pi_d^S G(a, b, 0)|_{S_a} = \left(b^2 \begin{pmatrix} 1 & 0 \\ 0 & 1 \end{pmatrix} + \begin{pmatrix} -b^2 & 0 \\ b^2 & 0 \end{pmatrix} \right) \begin{pmatrix} 0 \\ -b \end{pmatrix}.$$

Since S_a is a graph over the b-coordinate chart, we study the desingularised problem in this single coordinate chart and obtain

$$\dot{b} = -b^3 ; \tag{5.21}$$

compare with the corresponding reduced equation (2.59). Hence, the reduced flow along S_a is towards the origin where the two branches of the critical manifold cross, i.e., towards $S_a \cap S_d$.

Blow-Up Along S_d:
Since $Df|_{S_d} = (0, 0)$ is the zero vector, we are dealing with a line of nilpotent equilibria along S_d. Note further that this implies that the (desingularised) projector Π_d^S (4.20) onto S_d is not well defined. The mathematical tool that unravels nontrivial dynamics near nilpotent equilibria is known as the *blow-up* method.

Remark 5.6 *For an introduction to the blow-up method for nilpotent singularities see, e.g., Dumortier [27] and Bruno [16].*

In the case of the autocatalator model, we are interested in the dynamics localised near the line of nilpotent equilibria $b = 0$, i.e., the manifold S_d. We apply this technique to system (2.57) evolving on the fast time scale extended by the trivial equation $\dot{\varepsilon} = 0$, i.e., to the following three-dimensional system

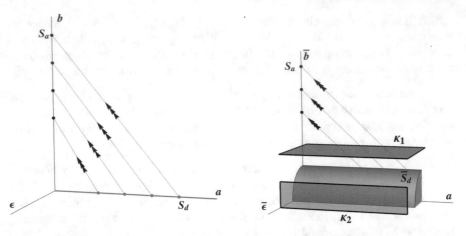

Fig. 5.6 Cylindrical blow-up of the line of degenerate equilibria S_d to a cylinder $\bar{S}_d$. The blown-up vector filed is studied in two coordinate charts $\kappa_1 : \bar{b} = 1$ and $\kappa_2 : \bar{\varepsilon} = 1$

$$\varepsilon \dot{a} = -ab^2 - \varepsilon^2 a + \varepsilon^2 \mu$$
$$\varepsilon \dot{b} = ab^2 - \varepsilon b + \varepsilon^2 a \qquad (5.22)$$
$$\dot{\varepsilon} = 0 \,.$$

This extension of the phase space by including the singular perturbation parameter as an additional variable is necessary because we are dealing with nilpotent equilibria in a singular perturbation problem. Here, the specific 'blow-up' of the line S_d is defined as the ε-dependent transformation

$$b = \bar{r}\bar{b}$$
$$\varepsilon = \bar{r}\bar{\varepsilon},$$

where $\bar{b}^2 + \bar{\varepsilon}^2 = 1$ is a compact manifold, a circle $\mathbb{S}^1$. Note that the variable a does not take part in this transformation. Hence, the line S_d is blown up to a cylinder $\mathbb{S}^1 \times \mathbb{R}$ and we refer to this transformation as a *cylindrical blow-up*[7] of the line S_d; see Fig. 5.6. The corresponding blown-up vector field is usually studied in appropriate coordinate charts of the blown-up cylindrical manifold. Here, two charts suffice: chart κ_1 defined by setting $\bar{b} = 1$, and chart κ_2 defined by setting $\bar{\varepsilon} = 1$.

---

[7] If this technique is applied to an isolated nilpotent singularity then it is usually referred to as a *spherical* blow-up since all variables participate in the blow-up.

The system in chart κ_2 with local coordinates[8] $(a_2, b_2, r_2) = (a, c, \varepsilon)$ is nothing else but the rescaled (zoomed) system (2.60) already discussed in Sect. 2.2.4. For $\varepsilon = 0$, it describes the fast flow on the blown-up cylinder shown in Fig. 5.6 (right) away from $\bar{b} = 0$. Recall the corresponding critical manifold $S^z = S_a^z \cup F^z \cup S_r^z$ (2.61) of this singularly perturbed system in standard form shown in Fig. 2.13 which consists of an attracting branch S_a^z, a repelling branch S_r^z and a fold point F^z. The corresponding reduced (slow) flow on the critical manifold S^z is also indicated in Fig. 2.13 as well as the unstable equilibrium point (2.65) on S_r^z for $\mu > 1$; see page 33 for details.

To understand how the dynamics in chart κ_2 connect with the super-fast/fast dynamics away from $b = 0$ shown in Fig. 5.6 (left), we need to look into chart κ_1 (i.e., $\bar{b} = 1$) with local coordinates $(a_1, r_1, \varepsilon_1) = (a, b, \varepsilon/b)$. The corresponding system is given by

$$\varepsilon_1 \dot{a} = -r_1(a + \varepsilon_1^2 a - \varepsilon_1^2 \mu)$$
$$\varepsilon_1 \dot{r}_1 = r_1(a - \varepsilon_1 + \varepsilon_1^2 a) \tag{5.23}$$
$$\dot{\varepsilon}_1 = a - \varepsilon_1 + \varepsilon_1^2 a;$$

note that $a = a_1$. Neither r_1 nor ε_1 are parameters in this coordinate chart, but they are variables that obey $r_1 \dot{\varepsilon}_1 + \varepsilon_1 \dot{r}_1 = 0$ which follows from the definition $r_1 \varepsilon_1 = \varepsilon$, i.e., the fibres $\varepsilon = const$ are bended in the blown-up system. To better understand the dynamics in chart κ_1, we rescale (the fast) time by $d\tau = \varepsilon_1 dt_1$ which gives the equivalent system

$$\dot{a} = -r_1(a + \varepsilon_1^2 a - \varepsilon_1^2 \mu)$$
$$\dot{r}_1 = r_1(a - \varepsilon_1 + \varepsilon_1^2 a) \tag{5.24}$$
$$\dot{\varepsilon}_1 = \varepsilon_1(a - \varepsilon_1 + \varepsilon_1^2 a),$$

where with a slight abuse of notation $\dot{} = d/dt_1$. System (5.24) has two invariant two-dimensional sets $\{r_1 = 0\}$ and $\{\varepsilon_1 = 0\}$ which correspond to the two branches of the blown-up fibre $\{\varepsilon = 0\}$, i.e., the blown-up cylinder $\mathbb{S}^1 \times \{\bar{r} = 0\}$ and the flat sheet $\{\bar{\varepsilon} = 0\}$.

On the cylinder $r_1 = 0$, we recover the singularly perturbed system in standard form of chart κ_2, i.e.,

$$\dot{a} = 0$$
$$\dot{\varepsilon}_1 = \varepsilon_1(a - \varepsilon_1 + \varepsilon_1^2 a), \tag{5.25}$$

with the critical manifold S^z given as a graph

$$a(\varepsilon_1) = \frac{\varepsilon_1}{\varepsilon_1^2 + 1}. \tag{5.26}$$

---

[8] Blow-up coordinates are denoted in each coordinate chart κ_i by a subscript i; see also Remark 3.2.

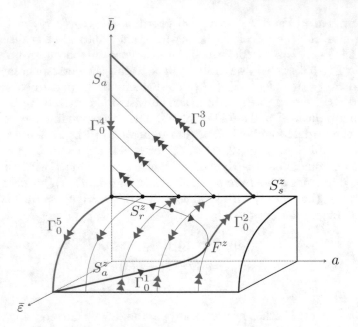

Fig. 5.7 The blown-up singular relaxation cycle $\Gamma_0 = \cup_i \Gamma_0^i$, $i = 1, \ldots, 5$, of the autocatalator model for $\mu > 1$ consists of one slow segment Γ_0^1 (blue), three fast segments Γ_0^2, Γ_0^4, Γ_0^5 (magenta) and one super-fast segment Γ_0^3 (red)

The critical manifold $S^z = S_a^z \cup F^z \cup S_r^z$ consists of an attracting branch S_a^z for $\varepsilon_1 > 1$, a repelling branch S_r^z for $0 < \varepsilon_1 < 1$ and a fold F^z for $\varepsilon_1 = 1$. Note that the critical manifold (5.26) of chart κ_1 and the critical manifold (2.61) of chart κ_2 are indeed the same since $\varepsilon_1 = 1/b_2$ defines the transition map κ_{21} from chart κ_1 to κ_2. In chart κ_1 we see that the unbounded branch S_r^z from chart κ_2, connects to the origin in κ_1; see Fig. 5.7.

On the invariant subspace $\varepsilon_1 = 0$, we recover the ultra-fast (layer) dynamics

$$
\begin{aligned}
\dot{a} &= -r_1 a \\
\dot{r}_1 &= r_1 a,
\end{aligned}
\tag{5.27}
$$

as shown in Fig. 5.6 (left). The ultra-fast dynamics stems from the time transformation $d\tau = \varepsilon_1 dt_1$ which identifies t_1 as significantly faster than τ as $\varepsilon_1 \to 0$ (i.e., ultra-fast). We have already analysed the ultra-fast and fast dynamics away from $r_1 = 0$.

An important (and final) piece of our dynamic puzzle is the one-dimensional intersection of the two invariant subspaces, $\{\varepsilon_1 = 0\} \cap \{r_1 = 0\}$—it is a line of equilibria, denoted by S_s^z, which is normally hyperbolic for $a \neq 0$; it is of saddle type with one positive and one negative real eigenvalue. It is this saddle structure of the line S_s^z that connects the fast dynamics on the cylinder (5.25) with the ultra-fast dynamics away from the cylinder (5.27) as can be seen in Fig. 5.7. In particular, we are now able to construct a singular relaxation cycle $\Gamma_0 = \cup_i \Gamma_0^i$, $i = 1, \ldots 5$, as a concatenation of orbit segments of three limiting systems reflecting the three different time scales observed. Gucwa and Szmolyan [36] show the existence of a nearby relaxation oscillator[9] for sufficiently small perturbations $0 < \varepsilon \ll 1$; for details, we refer the interested reader to this excellent paper [36].

---

[9] To prove this result rigorously, one needs to perform two additional spherical blow-ups of the two nilpotent equilibria (on the blown-up cylinder): the non-hyperbolic origin and the fold F^z; see [36] for details.

Chapter 6
Pseudo Singularities and Canards

There are important questions in the context of relaxation oscillatory behaviour observed in singularly perturbed systems that we have not addressed so far:

- where do these oscillations originate from, i.e., what is the underlying mechanism that leads to the genesis of rhythmic behaviour?
- how does the singular nature of the perturbation problem manifest itself in the rhythm generation?

Partial answers to the above questions can be found in classic *bifurcation theory* [63] which focuses on understanding significant changes in dynamical systems outputs under system parameter variations. The time-scale splitting in our singular perturbation problems creates additional complexity and sometimes surprising, counter-intuitive behaviour. We start with a couple of examples to motivate the development of the corresponding theory.

Recall the vdP relaxation oscillator system (2.22). We introduce an additional system parameter $\alpha \in \mathbb{R}$ to this singular perturbation problem in standard form,

$$x' = -\varepsilon(y - \alpha)$$
$$y' = x + y - \frac{y^3}{3} . \tag{6.1}$$

This parameter $\alpha \in \mathbb{R}$ represents a (dimensionless) applied current to this electric circuit model which could be either excitatory ($\alpha > 0$) or inhibitory ($\alpha < 0$). The corresponding bifurcation diagram with α as bifurcation parameter shown in Fig. 6.1 reveals that system (6.1) is in an excitable state for sufficiently strong inhibitory current $\alpha < -1$, i.e., there exists a stable rest state, and that the transition from an excitable state to relaxation oscillations happens near $\alpha = -1$ via a (supercritical) Andronov–Hopf bifurcation

M. Wechselberger, *Geometric Singular Perturbation Theory Beyond the Standard Form*, Frontiers in Applied Dynamical Systems: Reviews and Tutorials 4, https://doi.org/10.1007/978-3-030-36399-4_6

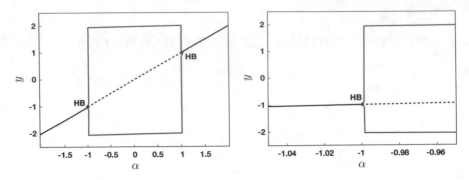

Fig. 6.1 (Left) bifurcation diagram of the forced vdP model (6.1) for $\varepsilon = 0.01$: there are two supercritical (singular) Andronov–Hopf bifurcations at $\alpha = \pm 1$; (Right) zoom that shows the supercritical (singular) Andronov–Hopf bifurcation (HB) at $\alpha = -1$ followed by a 'canard explosion' that terminates in the relaxation cycle branch

followed by a dramatic, steep incline of the oscillator amplitude that settles to the 'large' amplitude of the observed relaxation oscillations for $-1 < \alpha < 1$. In the singular limit $\varepsilon \to 0$, this steep increase in amplitude becomes vertical which indicates a *singular* nature of the bifurcation problem as well, i.e., it cannot be explained by classic bifurcation theory alone.

The same qualitative behaviour can be found in the autocatalator model (2.54) with a focus on the bifurcation parameter $\mu \geq 0$.[1] As discussed in Sect. 5.3, for $\mu < 1$ the autocatalator model is in an excitable state, i.e., the system rests at its equilibrium state, while for $\mu > 1$ it is in an (relaxation) oscillatory state. The corresponding bifurcation diagram with μ as bifurcation parameter shown in Fig. 6.2 shows the same qualitative features as in the forced vdP case: the transition from an excitable state to relaxation oscillations happens via a (supercritical) *singular* Andronov–Hopf bifurcation followed by a dramatic, steep incline of the oscillator amplitude near $\mu = 1$ that settles to the 'large' amplitude of the observed relaxation oscillations for $\mu > 1$.

Part of the explanation for the observed singular bifurcation structure is closely related to the violation of the regular jump point condition (4.22) along the set of contact points F.

Definition 6.1 *Any contact point* $z \in F$, *where*

$$N \, adj \, (DfN) DfG(z, 0) = \mathbb{O}_n \qquad (6.2)$$

is called a pseudo singularity. *We denote the set of pseudo singularities by*

---

[1] Variations in the dimensionless parameter μ could be interpreted as variations in the initial precursor concentration p_0.

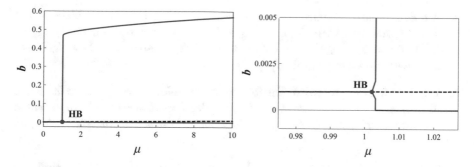

Fig. 6.2 (Left) Bifurcation diagram of the autocatalator model (2.54) for $\varepsilon = 0.001$; (Right) zoom that shows a (supercritical) singular Andronov–Hopf bifurcation (HB) at $\mu = 1.002$ followed by a canard explosion that terminates in the relaxation cycle branch

$$M_f := \{z \in F \; : \; N\,adj\,(DfN)DfG(z,0) = \mathbb{O}_n\}. \tag{6.3}$$

Remark 6.1 *Condition* (6.2) *is a single defining equation since the rank of* $adj\,(DfN)$ *is one along* F; *see Remark 4.2. Consequently, there is a marked difference between the* $k = 1$ *case (see Sect. 6.1) and the* $k > 1$ *case (see Sect. 6.3).*

Pseudo singularities[2] are equilibria of the desingularised problem (4.20) but not necessarily of the corresponding reduced problem (4.19) itself. At such a pseudo singularity $z_{ps} \in M_f \subseteq F$, the vector field of the reduced problem (4.19) contains indeterminate form(s) which could prevent a finite time blow-up of certain solutions of the reduced problem (4.19) approaching a pseudo singularity $z_{ps} \in M_f$. Such special solutions of the reduced problem that are able to pass in finite time through such a pseudo singularity from one branch of S to another are called *canards*,[3] and they play an important role in understanding the genesis of relaxation oscillations as well as in creating more complex oscillatory patterns in singular perturbation problems.

Definition 6.2 *Given a singularly perturbed system* (3.15) *with a* k-*dimensional critical manifold* S *that has contact (of order one) with the corresponding layer flow along a* $(k-1)$-*dimensional submanifold* $F \subset S$.

- *A singular canard* γ_c *is a trajectory of the reduced problem* (4.19) *that crosses with finite (nonzero) speed from an attracting normally hyperbolic branch* S_a *of the critical manifold* S *to a 'repelling' branch* $S_{r/s}$ *via a pseudo singularity* $z_{ps} \in M_f$ (6.3).

---

[2] The terminology 'pseudo singularity' was introduced by Jose Argemi [2].

[3] The terminology 'canard' was introduced by Eric Benoit et al. [8]; see also Remark 6.5.

- A singular faux[4] canard γ_f is a trajectory of the reduced problem (4.19) that crosses with finite (nonzero) speed from a 'repelling' normally hyperbolic branch $S_{r/s}$ of the critical manifold S to an attracting branch S_a via a pseudo singularity $z_{ps} \in M_f$ (6.3).

Remark 6.2 *This canard definition could be relaxed to problems where a trajectory crosses from any type of normally hyperbolic branch of S via a pseudo singularity $z_{ps} \in M_f \subseteq F$ to another normally hyperbolic branch of S.*

In the case of the forced vdP relaxation oscillator model (6.1), we have the same cubic-shaped critical manifold $S = \{(x, y) \in \mathbb{R}^2 : x = y^3/3 - y\}$ as in the unforced case ($\alpha = 0$), i.e., $S = S_a^- \cup F^- \cup S_r \cup F^+ \cup S_a^+$, with two regular contact points $F^\pm = \{(x, y) \in S : y = \pm 1\}$ for $|\alpha| \neq 1$. On the other hand, condition (6.2) is fulfilled for $\alpha = \pm 1$. Thus system possesses a pseudo singularity for those specific parameter values,

$$M_f^\pm = \{(x, y) \in F^\pm : \alpha = \pm 1\}.$$

The reduced problem can be solely studied in the y-coordinate chart and is given by

$$\dot{y} = \frac{y - \alpha}{1 - y^2}. \tag{6.4}$$

We have an equilibrium $y_{eq} = \alpha$, $\alpha \neq \pm 1$, in the reduced problem, which is unstable for $|\alpha| < 1$ and located on the repelling middle branch S_r, or it is stable for $|\alpha| > 1$ located on one of the outer two attracting branches $S_a^\pm$. For $\alpha = \pm 1$, there exists no equilibrium in the reduced problem due to the cancellation of a simple zero, i.e.,

$$\dot{y} = \frac{\mp 1}{1 \pm y}, \qquad \text{for } \alpha = \pm 1. \tag{6.5}$$

Thus we identify for $\alpha = -1$ a singular canard γ_c^-, i.e., a solution of (6.5) crossing from S_a^- via the pseudo singularity M_f^- to S_r and terminating at the regular jump point F^+. We identify for $\alpha = +1$ another singular canard γ_c^+, i.e., a solution of (6.5) crossing from S_a^+ via the pseudo singularity M_f^+ to S_r and terminating at the regular jump point F^-.

The existence of singular canards provides the possibility to define the so-called *singular canard cycles*, i.e., the concatenation of singular canard segments of the reduced problem with nonlinear fast fibre segments of the layer problem locally near the corresponding pseudo singularity to form singular cycles.

---

[4] French: 'faux' = false as opposed to 'vrai' = true; the term 'vrai' is seldom used in the canard theory literature.

Definition 6.3 *A singular canard cycle Γ_c is a concatenation of orbit segments of the reduced problem including (at least) one canard segment and (at least) one fast nonlinear fibre segment that form a closed loop.*

While singular relaxation cycles are, in general, unique, singular canard cycles are, in general, not, i.e., they show up as a whole family of singular canard cycles. For a given singular canard there exists locally a family of layer problem segments that connect this singular canard on adjacent branches S_a and $S_{r/s}$ near the corresponding pseudo singularity $z_{ps} \in M_f \subseteq F$. This is a consequence of the (order one) contact of the layer flow and the critical manifold S along F including the pseudo singularity $z_{ps} \in M_f$. Figure 6.3 shows such a family of singular canard cycles for $\alpha = -1$ in the forced vdP oscillator.

Canard theory, as part of geometric singular perturbation theory, is concerned with the following question:

- Do singular canards/canard cycles persist as canards/canard cycles for $\varepsilon > 0$?

To answer this question, we first have to provide a proper definition of a canard solution for $\varepsilon > 0$. Recall that the branches S_a and $S_{s/r}$ of the critical manifold S are normally hyperbolic (away from the set of contact points $F \subset S$ where normal hyperbolicity is lost). Fenichel theory implies the existence of a (non-unique but exponentially close) invariant slow manifold $S_{a,\varepsilon}$ and a (non-unique but exponentially close) invariant slow manifold $S_{s/r,\varepsilon}$ away from F; see Sect. 3.8 for details. Fix a representative for each of these manifolds.

Definition 6.4 *A maximal canard corresponds to the intersection of the manifolds $S_{a,\varepsilon}$ and $S_{s/r,\varepsilon}$ extended by the flow of (3.15) into a neighbourhood of the set of pseudo singularities $M_f \subseteq F$.*

Remark 6.3 *Such a maximal canard defines a family of canards nearby which are exponentially close to the maximal canard—a consequence of the non-uniqueness of $S_{a,\varepsilon}$ and $S_{s/r,\varepsilon}$. In the singular limit $\varepsilon \to 0$, such a family of canards is represented by a unique singular canard.*

6.1 Canard Theory: The Case $k = 1$

Here, the set of pseudo singularities $M_f \subseteq F$ (6.3) is degenerate because the set of contact points F itself consists only of isolated points that do not, in general, fulfil simultaneously the pseudo singularity condition (6.2). Thus in the case $k = 1$, a pseudo singularity has to be a codimension-one phenomenon, i.e., it is only observed under the variation of an additional system parameter, say $\alpha \in \mathbb{R}$, in system (3.10).[5]

---

[5] More generally, it is observed under variation along a path $\alpha(s)$, $s \in I \subset \mathbb{R}$, in the corresponding system parameter space $\alpha \in \mathbb{R}^p$, $p \geq 1$.

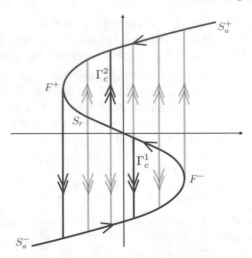

Fig. 6.3 A family of singular canard cycles in the forced van der Pol oscillator model (6.1) with $\alpha = -1$. There are two types of canard cycles, Γ_c^1 and Γ_c^2 within this family. Γ_c^1 cycles consist of two segments: one canard segment from S_a^- to S_r and one fast (jump back) segment from S_r to S_a^-. Γ_c^2 cycles consist of four segments: one canard segment from S_a^- to S_r, one fast (jump forward) segment from S_r to S_a^+, a reduced flow segment on S_a^+ and a fast segment from F^+ to S_a^-

Note that for $k = 1$, the rank of the *projection* operator Π^S in (4.19), respectively, Π_d^S in (4.20) is one, independent of $z \in S$ (i.e., including $z \in F \subset S$). Hence, condition (4.21) represents a single defining equation for an isolated singularity (equilibrium) $z_{eq} \in S \backslash F$ as does condition (6.2) for the existence of a pseudo singularity $z_{ps} \in F$. Assuming continuous dependence on the above mentioned additional system parameter $\alpha \in \mathbb{R}$ in the desingularised problem (4.20) and the existence of a pseudo singularity for $\alpha = \alpha_c$, there exists necessarily a relationship between an isolated contact point $F(\alpha)$ and an isolated singularity (equilibrium) $z_{eq}(\alpha)$ under the variation of $\alpha \in \mathbb{R}$. In general, one assumes that such an equilibrium $z_{eq}(\alpha)$ of the desingularised problem (4.20) crosses at $\alpha = \alpha_c$ from one normally hyperbolic branch of S to another via a pseudo singularity $F(\alpha_c) = M_f \subset S$ under the variation of $\alpha \in \mathbb{R}$, i.e., $F(\alpha_c) = z_{ps}(\alpha_c)$ in (4.20).

It is important to notice that for $\alpha = \alpha_c$ and for the corresponding pseudo singularity $F(\alpha_c) = M_f$ which is assumed to be a contact point of order one, there exists actually no singularity or equilibrium of the reduced problem (4.19) at $z = F(\alpha_c) = M_f$ due to the cancellation of a 'simple zero' in

$$\Pi^S G(z,0) = \frac{(-\det(DfN)\mathbb{I}_n + N \operatorname{adj}(DfN)Df)G(z,0)}{-\det(DfN)}, \qquad (6.6)$$

i.e., solutions of the reduced problem (4.19) are able to cross with finite speed from one branch of S via the pseudo singularity $F(\alpha_c) = M_f$ to the adjacent branch. These are the aforementioned singular canards (or singular faux canards); see Definition 6.2.

6.1.1 Singular Andronov–Hopf Bifurcation and Canard Explosion

For $k = 1$, the existence of a pseudo singularity M_f is related to the crossing of an equilibrium of the reduced (respectively, desingularised) problem under the variation of a system parameter $\alpha \in \mathbb{R}$. This indicates a bifurcation of the equilibrium state in the full system (3.10) under the variation of this parameter α. Two eigenvalues change simultaneously sign as the equilibrium crosses through the pseudo singularity $M_f = F'(\alpha_c)$—one related to the stability change in the reduced problem and the other related to the stability change in the layer problem. This simultaneous change of real eigenvalues in the two limiting problems is, rather surprisingly, an indicator of the crossing of a complex-conjugate pair of eigenvalues in the full system (3.10) for $\varepsilon \neq 0$ and, hence, we expect to observe an Andronov–Hopf bifurcation under the variation of the parameter $\alpha \in \mathbb{R}$. To identify and confirm this Andronov–Hopf bifurcation, we look at a corresponding planar system

$$\tilde{z}' = \tilde{H}(\tilde{z}, \varepsilon, \alpha) = \tilde{N}(\tilde{z}, \alpha)\tilde{f}(\tilde{z}, \alpha) + \varepsilon\tilde{G}(\tilde{z}, \alpha, \varepsilon), \tag{6.7}$$

where $\tilde{z} \in \mathbb{R}^2$ represent the local coordinates that describe the flow on the two-dimensional centre manifold W^c near such a pseudo singularity $\tilde{z}_{ps}$; see Sect. 4.4 for details on such a local centre manifold reduction.

Theorem 6.1 *Given a planar singularly perturbed system* (6.7) *with one-dimensional critical manifold* $S = S_a \cup F \cup S_r$ *that loses normal hyperbolicity at an order one contact point* F. *Assume further there exists a pseudo singularity* $\tilde{z}_{ps} = \tilde{z}_{ps}(\alpha_{ps})$ *for* $\alpha = \alpha_{ps} \in \mathbb{R}$ *and*

$$(\mathrm{tr}(adj(\tilde{N}D\tilde{f})D\tilde{G}) - \mathrm{tr}(adj(\tilde{G}D\tilde{f})D\tilde{N}))|_{\tilde{z}_{ps}} > 0. \tag{6.8}$$

Then a singular Andronov–Hopf bifurcation is observed in (6.7) *for* $\alpha = \alpha_{ah}(\varepsilon)$ *that creates small* $O(\sqrt{\varepsilon})$ *amplitude limit cycles with nonzero frequencies of order* $O(\sqrt{\varepsilon})$.

Proof An isolated equilibrium $\tilde{z}_{eq} \in \mathbb{R}^2$ of system (6.7) fulfils

$$\tilde{N}(\tilde{z}_{eq}, \alpha)\tilde{f}(\tilde{z}_{eq}, \alpha) = -\varepsilon\tilde{G}(\tilde{z}_{eq}, \alpha, \varepsilon). \tag{6.9}$$

It is assumed that this equilibrium corresponds in the limit $\varepsilon \to 0$ to an equilibrium of the one-dimensional critical manifold S, i.e., it is also an equilibrium of the reduced problem. The Jacobian of the planar system (6.7) is given by

$$D\tilde{H} = D\tilde{N}\tilde{f} + \tilde{N}D\tilde{f} + \varepsilon D\tilde{G}, \tag{6.10}$$

and the corresponding trace and determinant of the Jacobian by

$$\operatorname{tr} D\tilde{H} = \tilde{f}\operatorname{tr} D\tilde{N} + D\tilde{f}\tilde{N} + \varepsilon\operatorname{tr} D\tilde{G}$$

$$\det D\tilde{H} = \tilde{f}^2 \det D\tilde{N} + \varepsilon^2 \det D\tilde{G} + \varepsilon\tilde{f}\operatorname{tr}(\operatorname{adj}(D\tilde{N})D\tilde{G}) \tag{6.11}$$

$$+ \varepsilon(\operatorname{tr}(\operatorname{adj}(\tilde{N}D\tilde{f})D\tilde{G}) - \operatorname{tr}(\operatorname{adj}(\tilde{G}D\tilde{f})D\tilde{N})).$$

Remark 6.4 *We recall some helpful identities for $n \times n$ matrices A, B, such as*

$$\det(\alpha A) = \alpha^n \det A,$$

$$\operatorname{adj}(\alpha A) = \alpha^{n-1} \operatorname{adj} A, \tag{6.12}$$

$$D_t \det(A) = \operatorname{tr}(\operatorname{adj}(A)D_t A),$$

with $\alpha, t \in \mathbb{R}$, while the following formula is only valid for 2×2 matrices A, B:

$$\det(A + B) = \det A + \det B + \operatorname{tr}(\operatorname{adj}(A)B). \tag{6.13}$$

Under the variation of the system parameter α in system (6.7), an Andronov–Hopf bifurcation is observed for $\alpha = \alpha_{ah}(\varepsilon)$, $0 < \varepsilon \ll 1$, when $\operatorname{tr} D\tilde{H} = 0$ and $\det D\tilde{H} > 0$. The trace condition gives

$$D\tilde{f}\tilde{N} = -\tilde{f}\operatorname{tr} D\tilde{N} - \varepsilon\operatorname{tr} D\tilde{G}, \tag{6.14}$$

which implies that $D\tilde{f}\tilde{N} \to 0$ as $\varepsilon \to 0$, i.e., the singular bifurcation has to be located at a contact point F which confirms the definition of a pseudo singularity $\tilde{z}_{ps} = M_f = F$ in the case $k = 1$.

A necessary condition for the existence of a singular canard, i.e., a solution of the reduced problem crossing from S_a to S_r with finite speed, is that the equilibrium $\tilde{z}_{eq}$ is stable on S_a (respectively, unstable on S_r). For the planar system (6.7), the corresponding desingularised problem is given by

$$\Pi_d^S \tilde{G} = -\operatorname{adj}(\tilde{N}D\tilde{f})\tilde{G} = -\det(\tilde{N}|\tilde{G})(D\tilde{f}^\perp)^\top, \tag{6.15}$$

where $(\tilde{N}|\tilde{G})$ denotes a square matrix with column vectors $\tilde{N}$ and $\tilde{G}$. An equilibrium $\tilde{z}_{eq}$ is given by either $\det(\tilde{N}|\tilde{G})|_S = 0$ or $\tilde{G}|_S = \mathbb{O}_2$. The Jacobian of (6.15) evaluated at such an equilibrium is

$$\tilde{J}|_{\tilde{z}_{eq}} = -(D\tilde{f}^\perp)^\top D(\det(\tilde{N}|\tilde{G})), \tag{6.16}$$

and the stability condition for such an equilibrium $\tilde{z}_{eq}$ is then given by $\operatorname{tr} \tilde{J}|_{\tilde{z}_{eq}} < 0$ since $\det \tilde{J}|_{\tilde{z}_{eq}} = 0$, i.e., there is only one nontrivial eigenvalue. Using the matrix identities (6.12) and (6.13) gives the condition (6.8) which implies that $\det D\tilde{H} = O(\varepsilon) > 0$. Thus a *singular Andronov–Hopf bifur-*

cation is observed for $\alpha = \alpha_{ah}(\varepsilon)$ that creates small $O(\sqrt{\varepsilon})$ amplitude limit cycles with nonzero frequencies of order $O(\sqrt{\varepsilon})$, i.e., the singular nature of the Andronov–Hopf bifurcation is encoded in both, amplitude and frequency. $\square$

Note in Figs. 6.1 and 6.2 that the $O(\sqrt{\varepsilon})$-amplitude branch of limit cycles of the Andronov–Hopf bifurcation suddenly changes dramatically its size. This almost vertical branch marks the unfolding of the singular canard cycles within an exponentially small parameter interval of the corresponding bifurcation parameter (α in Fig. 6.1, μ in Fig. 6.2). This phenomenon is often referred to as a *canard explosion* [8, 28, 57] in standard singular perturbation problems. The following summarises these observations for general singular perturbation problems such as (6.7):

Theorem 6.2 *Given a planar singularly perturbed system* (6.7) *with one-dimensional critical manifold $S = S_a \cup F \cup S_r$ that loses normal hyperbolicity at an order one contact point F. Assume further there exists a pseudo singularity $\tilde{z}_{ps} = \tilde{z}_{ps}(\alpha_{ps})$ for $\alpha = \alpha_{ps} \in \mathbb{R}$ that also allows for the existence of a singular canard γ_c. Then a singular Andronov–Hopf bifurcation and a canard explosion occur at*

$$\alpha_{ah} = \alpha_{ps} + a_1\,\varepsilon + O(\varepsilon^{3/2}) \qquad and \qquad (6.17)$$

$$\alpha_c = \alpha_{ps} + (a_1 + a_2)\,\varepsilon + O(\varepsilon^{3/2}). \qquad (6.18)$$

The coefficients a_1 and a_2 can be calculated explicitly and, hence, the type of Andronov–Hopf bifurcation (super- or subcritical).

Proof The existence of a singular Andronov–Hopf bifurcation was shown in Theorem 6.1. For the proof of a canard explosion in standard singular perturbation problems see [28, 57], and in general singular perturbation problems see [45]. $\square$

6.1.2 Comparison with the Standard Case

Recall that the standard desingularised problem (4.20) is given by (4.23) and that the pseudo singularity condition (6.2) for a contact (fold) point $z \in F$ becomes

$$\text{adj}\,(D_y f) D_x f\, g(x, y, 0) = 0. \qquad (6.19)$$

In the case $k = 1$, we have $x \in \mathbb{R}$ and, hence, $D_x f$ is a $(n - 1)$-dimensional (column) vector. Recall further that at the contact point F, each row of $\text{adj}\,(D_y f)$ is either identical zero or it forms a left null vector of $D_y f$. It follows that $\text{adj}\,(D_y f) D_x f \neq 0$ is a non-vanishing (column) vector evaluated at F (because Df must have full rank along S including $F \subset S$). Thus in the case $k = 1$, (6.19) implies $g = 0$ for $z_{ps} = M_f = F$. In general, this condition can only be fulfilled in a one-parameter family of vector fields where

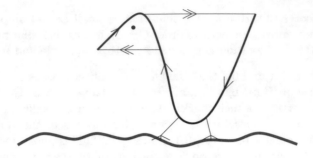

Fig. 6.4 A 'canard' (with head)

a singularity of the reduced problem, defined by $g = 0$, crosses the contact point F under the variation of a system parameter.

By Theorem 6.1, we identify a singular Andronov–Hopf bifurcation in the corresponding planar standard singular perturbation problem describing the flow on the two-dimensional centre manifold near such a pseudo singularity $z_{ps}(\alpha_{ps}) = (x_{ps}(\alpha_{ps}), y_{ps}(\alpha_{ps})) \in \mathbb{R}^2$ if

$$g(x, y, 0)|_{z_{ps}} = D_y f(x, y, 0)|_{z_{ps}} = 0 \quad \text{and} \quad D_x f(x, y, 0) D_y g(x, y, 0)|_{z_{ps}} < 0$$
$$(6.20)$$

are fulfilled. The above inequality is the standard equivalent to the inequality (6.8) which guarantees the existence of a singular canard. Such a singular Andronov–Hopf bifurcation is accompanied by the aforementioned canard explosion.

Remark 6.5 *Canards have been originally discovered by French mathematicians Eric Benoit et al. [8] who studied the vdP relaxation oscillator with constant forcing (6.1). With the canard phenomenon they were the first to explain the transition upon variation of the applied current $\alpha \in \mathbb{R}$ from small limit cycles via canard cycles to large amplitude relaxation cycles, and this transition happens in an exponentially small interval of the system parameter (canard explosion); see Fig. 6.1. In practice, only the small limit cycles or the large relaxation cycles are observed but not many of the canard cycles due to the exponentially sensitivity in the parameter. Even if the exponentially small parameter interval is known, numerical observation of all canard cycles with standard initial value solvers is impossible due to extreme sensitivity to numerical errors. Thus, their discovery seemed more like a hoax in a newspaper—a 'canard'. Furthermore, the canard cycles in the van der Pol oscillator resemble (with a little help of imagination) the shape of a 'duck'; see Fig. 6.4. Voilà, the notion of canard was born.*

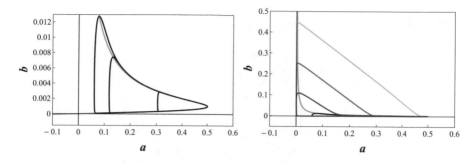

Fig. 6.5 *Canard explosion* in the autocatalator model observed in an exponentially small parameter window near $\mu_c \approx 1$: note the different b-scale in the two figures; (left) 'small' *jump-back* canards that only show slow-fast dynamics; (right)'large' *jump-forward* canards that show all three time scales involved—slow, fast and super-fast. Note, the largest canard cycle shown in the left panel is the same as the smallest canard cycle shown in the right panel

6.1.3 Canard Explosion in the Autocatalator Model

We revisit for the final time the autocatalator model (2.54). Recall that Fig. 6.2 shows the corresponding bifurcation diagram with μ as bifurcation parameter. Theorems 6.1 and 6.2 applied to the blown-up system (2.60) in chart κ_2 shows that the conditions in (6.20) are fulfilled, i.e., there exists a pseudo singularity $z_{ps}(\mu_{ps}) = (a_{ps}(\mu_{ps}), c_{ps}(\mu_{ps})) = (1/2, 1)$ with $\mu_{ps} = 1$ such that

$$D_a f\, D_c g|_{z_{ps}} = -(1 + c^2)|_{z_{ps}(\mu_{ps})} = -2 < 0\,,$$

and that the transition from an excitable state to relaxation oscillations happens near $\mu_{ps} = 1$ via a (supercritical) singular Andronov–Hopf bifurcation followed by a canard explosion.

While the bifurcation diagram shown in Fig. 6.2 and its steep canard cycle branch looks very similar to the one observed in the classic vdP relaxation oscillator[6] shown in Fig. 6.1, the geometry of the corresponding canard cycles is significantly different because of the underlying three time-scale structure of the autocatalator model. Figure 6.5 shows a few representatives of the family of canard cycles in this system. The left panel shows small 'jump-back' canard cycles that only involve two time scales (slow and fast), while the right panel shows 'jump-forward' canard cycles that involve all three time scales (slow, fast and super-fast).

This markedly different temporal behaviour is revealed when looking at the construction of the corresponding singular canard cycles shown in Fig. 6.6. The jump-back canard cycles consist only of two segments (one slow and one fast) like in the case of jump-back canards in the vdP relaxation oscillator,

---

[6] Theorems 6.1 and 6.2 apply to the forced vdP oscillator (6.1) as well.

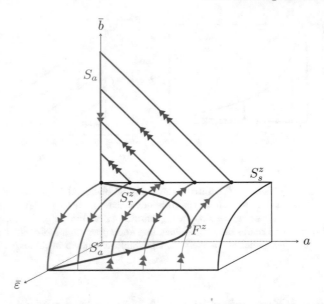

Fig. 6.6 The blown-up singular canard cycles of the autocatalator model for $\mu = 1$: *jump-back* canard cycles consist of one slow (blue) and one fast segment (magenta), while *jump-away* canard cycles consist of one slow (blue), three fast (magenta) and one super-fast segments (red)

while the jump-forward canard cycles consist of five segments (one slow, three fast and one super-fast) similarly to the singular relaxation cycle shown in Fig. 5.7.

The singular canard cycle that defines the boundary between jump-back and jump-forward canard cycles is given by the singular canard cycle that follows the repelling branch S_r^z all the way to the origin before jumping back to S_a^z (it corresponds to the maximal canard of this problem).

As mentioned before in the relaxation oscillator case, to rigorously prove the existence of the corresponding canard cycle explosion one needs to perform additional blow-ups at the two non-hyperbolic points at the origin and at the contact point F^z to prove existence. The proof for these canard cycles has been discussed in Ilona Kosiuk's thesis [52], Chapter 3.4.

Remark 6.6 *Similar canard explosion dynamics as in the autocatalator model (2.54) has been found in a model for aircraft ground dynamics by Rankin et al. [93]. Kristiansen [55] studied such models in a more general context of flat slow manifolds that lose normal hyperbolicity at infinity (as the autocatalator model studied in chart κ_2 and the above mentioned aircraft ground dynamics model), and so did Kuehn [60].*

6.1.4 A Variation of the Two-Stroke Oscillator

Let us consider the following variation of the two-stroke oscillator model (2.28),

$$x'' + \left(\frac{\varepsilon(\beta - \gamma x)}{\delta - x'} - \alpha x' \right) + x = 0 , \qquad (6.21)$$

with $' = d/d\tilde{t}$, singular perturbation parameter $\varepsilon \ll 1$, $0 < \alpha < 2$, $\beta > 0$, $\gamma \geq 0$ and $\delta > 0$ as additional system parameters.[7] For $\gamma \neq 0$, we are dealing with a displacement dependent damping model.[8] Similar to system (2.28), this extended model (6.21) can be recast as a dynamical system,

$$\begin{aligned} x' &= y \\ y' &= -x + \alpha y - \varepsilon \frac{(\beta - \gamma x)}{\delta - y} , \end{aligned} \qquad (6.22)$$

which is closely related to a singularly perturbed system in general form. This becomes evident by introducing a phase space dependent time-scale transformation,

$$d\tilde{t} = (\delta - y)dt , \qquad (6.23)$$

which is orientation preserving for $y < \delta$ and leads to

$$\begin{aligned} x' &= y(\delta - y) \\ y' &= (-x + \alpha y)(\delta - y) - \varepsilon(\beta - \gamma x) . \end{aligned} \qquad (6.24)$$

This system is of the general form (3.10) with

$$z = \begin{pmatrix} x \\ y \end{pmatrix} \in \mathbb{R}^2, \ N(x, y) = \begin{pmatrix} y \\ -x + \alpha y \end{pmatrix}, \ f(x, y) = \delta - y , \qquad (6.25)$$

and

$$G(x, y, \varepsilon) = \begin{pmatrix} 0 \\ -\beta + \gamma x \end{pmatrix} . \qquad (6.26)$$

In Fig. 6.7, the corresponding bifurcation diagram of system (6.24) with β as bifurcation parameter shows a (subcritical) singular Andronov–Hopf bifurcation and a corresponding (subcritical) canard explosion near $\beta_{ps} = 1$. Note, the unstable canard cycles lie all above $S_{a,\varepsilon}$ and $S_{r,\varepsilon}$, and ultimately terminate with 'infinite' amplitude. The bifurcation diagram also shows a second disconnected branch of stable relaxation and canard cycles (solid blue).

Layer Problem:
We note that the layer problem

---

[7] The original two-stroke oscillator model (2.28) is given by $\alpha = 1$, $\beta = 1$, $\gamma = 0$ and $\delta = 1$.
[8] Displacement dependent normal forces arise frequently in engineering problems [1, 26, 42, 114].

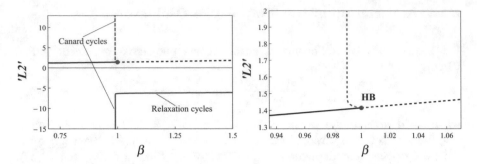

Fig. 6.7 Bifurcation diagram of the two-stroke oscillator model (6.21) with $\alpha = \gamma = \delta = 1$, $\varepsilon = 0.01$ and $\beta > 0$ as bifurcation parameter: (left) one branch of 'small' unstable canard cycles (dashed blue) emanating from a singular Hopf bifurcation (HB) and a second branch of 'large' stable canard cycles (solid blue) that is connected to the relaxation cycle branch. Figure 6.8 shows a representative of each canard cycle branch. The position of these limit cycles in phase space relative to their y-coordinate explains the signed 'L2'-norm used here to distinguish the different canard cycles (positive/negative). Note that both canard cycle branches are unbounded; (right) zoom that indicates a subcritical singular Hopf bifurcation for $\beta_{ah} \approx 1$

$$\begin{pmatrix} x' \\ y' \end{pmatrix} = \begin{pmatrix} y \\ -x + \alpha y \end{pmatrix} (\delta - y) \tag{6.27}$$

is almost identical to that of the original two-stroke oscillator model; see (2.37). We recall important facts for convenience. The set of singularities of (6.27) is given by

$$S_0 = \{y = \delta\} \cup \{(0,0)^\top\}.$$

This assumption guarantees that the isolated singularity at the origin is an unstable focus—it provides the necessary globally return mechanism through fast rotation. The subset

$$S = \{(x,y) \in \mathbb{R}^2 \ : \ y = \delta\} \subset S_0 \tag{6.28}$$

forms the one-dimensional critical manifold of this problem. The Jacobian Dh evaluated along $(x,y) \in S$ is given by

$$Dh|_S = \begin{pmatrix} 0 & -\delta \\ 0 & x - \alpha\delta, \end{pmatrix}$$

which has one trivial eigenvalue and $\lambda_1 = DfN|_S = x - \alpha\delta$, $\forall x \in \mathbb{R}$, as its nontrivial eigenvalue. Thus S loses normal hyperbolicity for $x = \alpha\delta > 0$, and $S = S_a \cup F \cup S_r$ consists of an attracting branch S_a for $x < \alpha\delta$, a repelling branch for $x > \alpha\delta$ and a single contact point

$$F = \{(x,y) \in S \ : \ x = \alpha\delta\}. \tag{6.29}$$

The conditions (4.6) of Lemma 4.1 are fulfilled, i.e.,

$$\mathrm{rk}\, Df|_F = n - k = 1\,,$$

$$l \cdot (D^2 f(Nr, Nr) + Df\, DN(Nr, r)) = Df\, DN\, N|_F = (0\ -1) \begin{pmatrix} 0 & 1 \\ -1 & \alpha \end{pmatrix} \begin{pmatrix} \delta \\ 0 \end{pmatrix} = \delta \neq 0\,,$$

which shows that the contact of the layer flow with S at F is of order one. Note further that the fast rotations around the isolated focal singularity $(0,0)^{\top}$ cause the order one contact of the layer flow with S at F.

Reduced Problem:
Here, we study the corresponding desingularised problem (4.20) evaluated along the critical manifold S (5.4) which is given by

$$\begin{pmatrix} \dot{x} \\ \dot{y} \end{pmatrix} = \left(-(x - \alpha\delta) \begin{pmatrix} 1 & 0 \\ 0 & 1 \end{pmatrix} + \begin{pmatrix} 0 & -\delta \\ 0 & x - \alpha\delta \end{pmatrix} \right) \begin{pmatrix} 0 \\ -\beta + \gamma x \end{pmatrix}.$$

S is a graph over the x-coordinate chart, and we study the desingularised flow in this single coordinate chart given by

$$\dot{x} = \delta(\beta - \gamma x)\,. \tag{6.30}$$

Since $\lambda_1 = \det(DfN)|_S = x - \alpha\delta$, the flow of this desingularised system has to be reversed on S_r, i.e., for $x > \alpha\delta$, to obtain the corresponding reduced flow of (4.19). For $\gamma > 0$, there exists a singularity of (6.30),

$$x = \frac{\beta}{\gamma} > 0\,. \tag{6.31}$$

For $0 < \beta < \alpha\gamma\delta$, it is a stable equilibrium of the corresponding reduced problem (4.19) on the attracting branch S_a while for $\beta > \alpha\gamma\delta > 0$, it is an unstable equilibrium of the corresponding reduced problem (4.19) on the repelling branch S_r.[9] By Definition 6.2, we have a single pseudo singularity

$$M_f = \{(x, y) \in F\ :\ Df\, G(x, y, 0) = 0\}\,, \tag{6.32}$$

for $\beta = \alpha\gamma\delta$ only. Note that for $\beta = \alpha\gamma\delta$ there is no equilibrium in the corresponding reduced problem (4.19) due to the cancellation of a simple zero, i.e., in this case the reduced flow is given by

$$\dot{x} = \gamma\delta > 0\,. \tag{6.33}$$

This allows for a singular canard to pass from S_a via M_f to S_r.

---

[9] For $\gamma = 0$, there is no equilibrium; compare with the original stick-slip oscillator model (2.28) analysis.

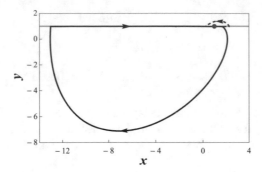

Fig. 6.8 Generalised two-stroke oscillator model (6.21) with $\alpha = \gamma = \delta = 1$ and $\varepsilon = 0.01$: a small unstable canard cycle (dashed) and a large stable canard cycle (solid) for $\beta = 0.990000000001$. The red dot is the single contact point F which indicates that both cycles consist of canard segments

Relaxation Oscillator:
Based on the above analysis, we have for $\beta > \alpha\gamma\delta$ an unstable equilibrium on S_r. For this parameter regime, we conclude the existence of a (unique) singular relaxation cycle $\Gamma = \Gamma_r \cup \Gamma_f$ which is a concatenation of an orbit segment Γ_r of the reduced problem connecting a point $p_l \in S_a$ at $x_l < \alpha\delta$ to the regular jump point $p_j = F$ at $x_j = \alpha\delta$ and a fast fibre segment Γ_f connecting $p_j = F$ back to $p_l \in S_a$; compare with Fig. 5.2.

By Proposition 5.1, such a singular relaxation cycle Γ perturbs to a nearby attracting relaxation cycle Γ_ε; compare with Fig. 2.6. For a detailed analysis, we refer the reader to [45].

Canard Cycles:
The above analysis also shows that for $\beta = \alpha\gamma\delta$ the reduced flow is able to cross from S_a via the pseudo singularity M_f (6.32) to S_r in finite time. Hence, we have identified a singular canard and we are able to construct (a family of) singular canard cycles that extend beyond $x = \alpha\delta > 0$ onto $S_{r,\varepsilon}$; see Fig. 6.8.

Theorems 6.1 and 6.2 apply and identify a (subcritical) singular Andronov–Hopf bifurcation and a corresponding (subcritical) canard explosion near $\beta_{ps} = 1$. Note that condition (6.8) are fulfilled, i.e., $z_{ps} = (\alpha\delta, \delta)$ for $\beta_{ps} = \alpha\gamma\delta$ and

$$(\text{tr}(\text{adj}\,(NDf)DG) - \text{tr}(\text{adj}\,(GDf)DN))|_{z_{ps}} = \text{tr}(\text{adj}\,(NDf)DG)|_{z_{ps}} = \delta\gamma > 0\,,$$

which confirms the reduced flow condition (6.33) for the existence of a singular canard. The corresponding bifurcation diagram is shown in Fig. 6.7 (right). Note, the unstable canard cycles lie all above $S_{a,\varepsilon}$ and $S_{r,\varepsilon}$, and ultimately terminate with 'infinite' amplitude.

The bifurcation diagram in Fig. 6.7 (left) also shows a second disconnected branch of relaxation and canard cycles (solid blue). The system is bistable for $\beta_c \leq \beta \leq \beta_{ah}$ because of the existence of stable relaxation cycles for $\beta > \beta_c$. These relaxation cycles lie all below $S_{a,\varepsilon}$ and $S_{r,\varepsilon}$. Near $\beta_{ps} = 1$, they undergo a 'canard explosion' and terminate with 'infinite' amplitude at $\beta = \beta_c$. The corresponding canard cycles are stable; see Fig. 6.8 where a stable large canard cycle from this family is shown together with a small unstable canard cycle from the canard cycle family born out from the singular Andronov–Hopf bifurcation.

Remark 6.7 *To study the termination of stable and unstable limit cycles with infinite amplitude as seen in Fig. 6.7, a compactification of the phase space is necessary.*

6.2 Excitability and Dynamic Parameter Variation

We turn our attention to the second question raised at the beginning of this chapter: how does the singular nature of the perturbation problem manifest itself in the rhythm generation? In particular, we are interested in rhythm generation when there is not necessarily a singular bifurcation involved. To motivate the development of the corresponding theory, we focus on the concept of excitability in the modified two-stroke oscillator model (6.21) in more detail.

Model Assumption 6.1 *In the two-stroke oscillator model* (6.21)*, we assume $0 < \alpha < 2$, $\gamma > 0$, $\delta > 0$ and $\beta < \alpha\gamma\delta$.*

As shown in Sect. 6.1.4, under the model assumption $\beta < \alpha\gamma\delta$ there exists a (global) stable equilibrium on $S_{a,\varepsilon}$, i.e., this system does not possess an oscillatory attractor state. On the other hand, this system might be able to respond to external inputs or significant changes in the other system parameters in 'large' transients, i.e., the system is able to show *excitability*.

Remark 6.8 *In living organisms, the term 'excitability' refers to the ability to respond strongly to the action of a relatively weak stimulus. A typical example of excitability is the generation of a spike (an action potential) in neurons induced by a short depolarising or hyperpolarising electrical perturbation of the resting membrane potential.*

For the two-stroke oscillator model (6.21), this excitability is illustrated in Fig. 6.9, where we discontinuously/quickly change the system parameter δ as indicated in the lower panel. In the upper panel we see the corresponding response of the excitable system. We only observe a 'small' response to the parameter change $\delta = 1 \rightarrow \delta = 1.2$ (blue), i.e., we observe that the system approaches monotonically the new rest state of the system. On the other

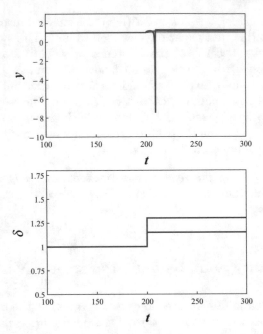

Fig. 6.9 Instantaneous change of the parameter δ (lower panel) causes either a transient response or not (upper panel) in system (6.21) which is in the excitable regime with $\alpha = 1.0$, $\beta = 0.5$, $\gamma = 1.0$ and $\varepsilon = 0.01$: the difference is the height of the instantaneous change of δ; blue curve, $\delta = 1 \rightarrow \delta = 1.15$; red curve, $\delta = 1 \rightarrow \delta = 1.3$

hand, we observe a 'large' response to the parameter change $\delta = 1 \rightarrow \delta = 1.4$ (red), i.e., the system makes a huge excursion, it spikes, before it approaches the new rest state of the system.

The dramatic difference between these two cases can be best explained by looking into the corresponding phase space (plane); see Fig. 6.10. Changing δ means changing the position of the critical manifold (6.28). If this critical manifold is (instantaneously) sufficiently raised then the initial condition of the altered system (which is the old resting state) lies on a fast 'fibre' of the altered system that is not anymore attracted locally to S_a and, hence, not locally to $S_{a,\varepsilon}$ for sufficiently small $\varepsilon > 0$, leading to a large excursion.

Remark 6.9 *Such discontinuous changes of parameters can be viewed as 'fast' changes or as a* control theory *problem. In neurophysiology, a current step protocol, i.e., a step increase or decrease of the applied current, is a classic procedure to test excitability of neurons.*

On the other hand, we may assume that system parameters change 'slowly' over time (seasonal influences, environmental changes, etc.). In the two-stroke oscillator model (6.21), we are interested in those 'slow' dynamic variations of, e.g., the parameter $\delta = \delta(\varepsilon \tilde{t})$, that may lead to a significant transient

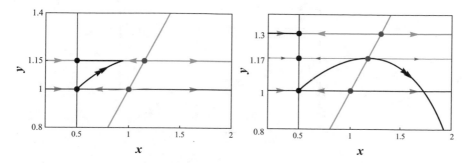

Fig. 6.10 Instantaneous change of the parameter delta δ causes either a transient response or not (upper panel) in system (6.21) depending on the height of the change: (Left) for $\delta = 1 \to \delta > 1.15$ the trajectory with IC at the old equilibrium $(x, y) = (0.5, 1)$ terminates on S_a along the fast fibre of the new base point, the solid black dot and then follows back to the now new equilibrium position at $(x, y) = (0.5, 1.15)$. (Right) for $\delta = 1 \to \delta > 1.3$ the trajectory with IC at the old equilibrium $(x, y) = (0.5, 1)$ passes S and makes a large excursion before returning to the now new equilibrium position at $(x, y) = (0.5, 1.3)$. The singular limit analysis shows that the parameter δ has to be instantaneously raised from $\delta = 1 \to \delta > 1.17$ to pass the contact point F along the fast fibre which confirms the observation made in Fig. 6.9

response, spike(s) or large oscillation(s). Figure 6.11 shows two examples for a parameter ramp given by

$$\delta(\varepsilon \tilde{t}) = \delta_- + \frac{\delta_1}{2}\left(1 + \tanh\left(\frac{2\varepsilon(\tilde{t} - \tilde{t}_0)}{s_1}\right)\right), \qquad (6.34)$$

with $\lim\limits_{\tilde{t} \to -\infty} \delta(\varepsilon \tilde{t}) = \delta_{min} = \delta_-$ and $\lim\limits_{\tilde{t} \to \infty} \delta(\varepsilon \tilde{t}) = \delta_{max} = \delta_- + \delta_1 =: \delta_+$ centred around $\tilde{t}_0$ where the ramp has its maximum slope of $\delta'_{max} = \varepsilon \delta_1 / s_1$. The height of the two ramps shown in Fig. 6.11 is the same, $\delta_+ = 1.4$. The difference is in the maximum slope of the ramps, δ'_{max}. In the case of the smaller maximum slope, we see a small response while in the other, larger maximum slope case, we see a large response, a spike. Clearly, the reason for the dramatic difference has to be the different temporal evolutions of the ramps since their height is the same.

To understand this observed transient phenomenon, we study the equivalent system (6.24),

$$\begin{aligned} x' &= y(\delta(\varepsilon \tilde{t}) - y) \\ y' &= (-x + \alpha y)(\delta(\varepsilon \tilde{t}) - y) - \varepsilon(\beta - \gamma x) \end{aligned} \qquad (6.35)$$

under the modification that $\delta = \delta(\varepsilon \tilde{t})$ is a slowly varying parameter.

Model Assumption 6.2 *In system* (6.35), *we fix* $0 < \alpha < 2$, $\gamma > 0$, $\delta > 0$. *The slowly varying parameter* $\delta = \delta(\varepsilon \tilde{t}) > 0$, $\forall \tilde{t} \in \mathbb{R}$, *is a bounded non-*

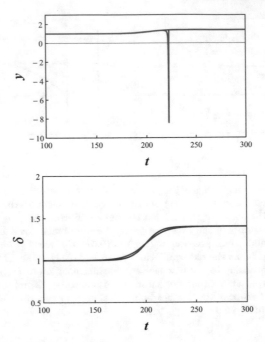

Fig. 6.11 (Lower panel) Two ramp protocols shown in red and blue for the slow modulation of the 'parameter' δ (6.34) with $\delta_- = 1$, $\delta_1 = 0.4$, $\tilde{t}_0 = 200$: the difference is the steepness of the ramp, or more precisely the temporal profile, encoded in the parameter $s_1 = 0.4$ (blue), $s_1 = 0.3$ (red), not the height of the ramp which is the same $\delta_1 = 0.4$. (Upper panel) The corresponding response of system (6.21) with $\alpha = 1.0$, $\beta = 0.5$, $\gamma = 1.0$ and $\varepsilon = 0.01$ for these ramp protocols: blue—none, red—spike

negative function which is asymptotically constant as $\tilde{t} \to \pm\infty$, i.e., $0 < \delta_{min} \le \delta(s) \le \delta_{max}$, $\lim\limits_{\tilde{t}\to\pm\infty} \delta(\varepsilon\tilde{t}) = \delta_\pm$ and $\lim\limits_{\tilde{t}\to\pm\infty} \delta'(\varepsilon\tilde{t}) = 0$. We also fix $0 < \beta < \alpha\gamma\delta_{max}$ to be in an excitable regime only.

Remark 6.10 *The ramp (6.34) fulfils this model assumption on $\delta(\varepsilon\tilde{t})$.*

We introduce a dummy slow variable $s := \varepsilon\tilde{t}$ to recast the nonautonomous two-dimensional system (6.35) as an autonomous three-dimensional system

$$
\begin{aligned}
s' &= \varepsilon \\
x' &= y(\delta(s) - y) \\
y' &= (-x + \alpha y)(\delta(s) - y) - \varepsilon(\beta - \gamma x),
\end{aligned}
\tag{6.36}
$$

which is of the general form (3.10) with

$$
z = \begin{pmatrix} s \\ x \\ y \end{pmatrix} \in \mathbb{R}^3, \quad N(s, x, y) = \begin{pmatrix} 0 \\ y \\ -x + \alpha y \end{pmatrix}, \quad f(s, x, y) = \delta(s) - y, \tag{6.37}
$$

and

$$G(s, x, y, \varepsilon) = \begin{pmatrix} 1 \\ 0 \\ -\beta + \gamma x \end{pmatrix}. \tag{6.38}$$

Note, the extended system (6.36) has no equilibria for $\varepsilon > 0$, because of the nonautonomous nature of the original problem (6.35). Thus no (singular) bifurcation will explain the observed excitability. On the other hand, we will show that the extended system (6.36) possesses pseudo singularities, and they provide the key to explain the observed dynamics which motivates the development of the following canard theory.

6.3 Canard Theory: The Case $k \geq 2$

Here, pseudo singularities are still singularities (equilibria) of the desingularised problem (4.20) but they are, in general, *not* necessarily related to equilibria of the reduced problem (4.19). Furthermore, they occur generically in such (higher dimensional slow manifold) problems, i.e., there is, in general, no need for the variation of an additional system parameter to observe them. This makes canards in the case $k \geq 2$ robust 'creatures', i.e., their impact on the dynamics of a singularly perturbed system is, in general, observable.

Definition 6.5 *For $k \geq 2$, the set of pseudo singularities M_f (6.2) forms a (union of) $(k-2)$-dimensional subset(s) of the critical manifold S.*

Remark 6.11 *In the following, M_f refers to one connected $(k-2)$-dimensional subset (a connected $(k-2)$-dimensional manifold) of pseudo singularities, unless otherwise stated.*

The set $M_f \subset F$ viewed as a set of equilibria of the desingularised problem (4.20) has generically $(k-2)$ zero eigenvalues and two nontrivial eigenvalues $\lambda_{1/2}$ with nonzero real part. Thus in the case $k > 2$, the existence of a $(k-2)$ dimensional set of pseudo singularities $M_f \subset F$ indicates an additional local time-scale splitting within the reduced problem (4.19) near $M_f \subset F \subset S$, i.e., the set M_f represents generically a $(k-2)$-dimensional normally hyperbolic manifold of pseudo singularities when viewed as a set of equilibria of the associated desingularised problem (4.20). This implies that a local invariant nonlinear foliation of S with base points $z_{ps} \in M_f$, denoted by $\mathcal{L}_s(M_f)$, is contained in a k-dimensional tubular neighbourhood $\mathcal{B}_s(M_f) \subset S$ of the $(k-2)$ dimensional invariant manifold M_f of the desingularised problem (4.20). Each two-dimensional nonlinear fibre of this locally invariant nonlinear foliation $\mathcal{L}_s(M_f) = \cup_{z_{ps} \in M_f} \mathcal{L}_s(z_{ps})$ is tangent to the eigenspace spanned by the two nontrivial eigenvalues of the corresponding pseudo singularity $z_{ps} \in M_f$. The classification of pseudo singularities is based on these two nontrivial eigenvalues $\lambda_{1/2}$ and follows that of singularities in two-dimensional vector fields.

Definition 6.6 *Given a pseudo singularity $z_{ps} \in M_f$ of the desingularised problem (4.20) with $k \geq 2$ and corresponding nontrivial eigenvalues $\lambda_{1/2} \in \mathbb{C}$.*

- *In the case that $\lambda_{1/2} \in \mathbb{R}$, let us denote the eigenvalue ratio by*

$$\mu := \lambda_1/\lambda_2 \in \mathbb{R}.$$

  The corresponding singularity $z_{ps} \in M_f$ is called a pseudo saddle for $\mu < 0$, or a pseudo node for $\mu > 0$.
- *In the case that $\lambda_{1/2} \in \mathbb{C}\backslash\mathbb{R}$ are complex conjugates with $\mathrm{Re}\,\lambda_{1/2} \neq 0$, the corresponding singularity $z_{ps} \in M_f$ is called a pseudo focus.*

For generic pseudo singularities, the algebraic multiplicity of the corresponding singularities on both, the right and left-hand sides in the reduced problem (4.19) is the same (which is one). This leads in the case of a pseudo saddle or a pseudo node to a nonzero but finite speed of the reduced flow through such a pseudo singularity $z_{ps} \in M_f \subset F$ in the eigendirection corresponding to the two nonzero eigenvalues. Hence, pseudo saddles and pseudo nodes create possibilities for the reduced flow to cross to different (normally hyperbolic) branches of the critical manifold S via such pseudo singularities $z_{ps} \in M_f \subset F$. Again, this is the hallmark of singular canards; see Definition 6.2.

Remark 6.12 *In the case $k = 2$, the set M_f consists of isolated pseudo singularities. This makes the description of associated geometric objects in the following often simpler or even trivial.*

Remark 6.13 *Eric Benoit [6] was the first who studied pseudo singularities and canards in the case $k = 2$.*

6.3.1 Pseudo Saddles

In the real eigenvalue ratio case $\mu < 0$, we split the nonlinear foliation $\mathcal{L}_s(M_f)$ of the $(k - 2)$-dimensional normally hyperbolic manifold of pseudo saddles M_f of the desingularised problem (4.20) into sub-foliations that reflect ('fast') motion towards or away from the set of base points $z_{ps} \in M_f$.

Definition 6.7 *The local stable and unstable manifolds of a normally hyperbolic manifold of pseudo saddles M_f, denoted by $W^s_{loc}(M_f)$ and $W^u_{loc}(M_f)$, respectively, are the (disjoint) unions*

$$W^s_{loc}(M_f) = \bigcup_{z_{ps} \in M_f} W^s_{loc}(z_{ps}), \quad W^u_{loc}(M_f) = \bigcup_{z_{ps} \in M_f} W^u_{loc}(z_{ps}), \quad (6.39)$$

where $W^s_{loc}(z_{ps})$ and $W^u_{loc}(z_{ps})$ are local stable and unstable manifolds of the base point $z_{ps} \in M_f$ of the desingularised problem (4.20). These sets

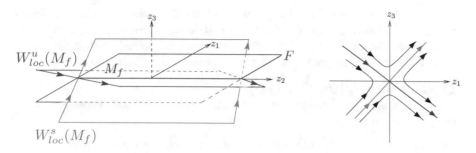

Fig. 6.12 Sketch of the local reduced flow ($k = 3$) near a local one-dimensional manifold M_f of pseudo saddles (red): (left) two-dimensional manifold of order one contact points F (blue), two-dimensional manifold $W^s_{loc}(M_f)$ (green) of singular canards through M_f and two-dimensional manifold $W^u_{loc}(M_f)$ (brown) of faux canards through M_f; (right) projection of reduced flow on a section $z_2 = $ constant

$W^s_{loc}(z_{ps})$ and $W^u_{loc}(z_{ps})$, respectively, form a family of locally invariant non-linear fibres for $W^s_{loc}(M_f)$ and $W^u_{loc}(M_f)$, respectively. The corresponding fibres are tangent to the one-dimensional stable eigendirection $E^s(z_{ps})$ and the one-dimensional unstable eigendirection $E^u(z_{ps})$, respectively, of each pseudo singularity $z_{ps} \in M_f$.

Remark 6.14 *The local manifolds $W^s_{loc}(M_f)$ and $W^u_{loc}(M_f)$, respectively, are connected and $(k-1)$-dimensional.*

In the reduced problem (4.19), singular canards of pseudo saddle type are trajectories that originate in a stable fibre $W^s_{loc}(z_{ps})$ on the stable branch S_a approach the set M_f in finite time and cross tangent to the stable eigendirection of the corresponding pseudo singularity $z_{ps} \in M_f$ to the 'unstable' branch $S_{r/s}$.

Remark 6.15 *All other trajectories of the reduced problem (4.19) originating on the stable branch S_a (close to F) reach either the set of regular jump points $F \backslash M_f$ in finite forward or backward time where they cease to exist due to finite time blow-up, or they do not reach the set $F \backslash M_f$ at all.*

In the reduced problem (4.19), singular faux canards of pseudo saddle type are trajectories starting in an unstable fibre $W^u_{loc}(z_{ps})$ on the 'unstable' branch $S_{r/s}$, approach the set M_f in finite time and cross tangent to the unstable eigendirection of the corresponding pseudo singularity $z_{ps} \in M_f$ to the stable branch S_a.

Remark 6.16 *In the case $k = 2$, we deal with isolated pseudo saddles $z_{ps} \in M_f$. The two eigendirections of the nontrivial eigenvalues of such a pseudo saddle z_{ps}, $W^s_{loc}(z_{ps})$ and $W^u_{loc}(z_{ps})$, respectively, correspond to a singular canard and a singular faux canard, respectively, in the reduced problem where solutions are able to cross with finite speed from one branch of S to another via $z_{ps} \in M_f$; compare with Fig. 6.12 (right).*

In the classic GSPT literature, pseudo singularities are referred to as *folded singularities* due to the local 'folded' geometric structure of the critical manifold S; see Definition 4.3. All standard results on folded singularities and associated canards apply via local equivalence to the general case.

Theorem 6.3 (cf. [80, 102, 111])

In the pseudo saddle case $\mu < 0$ of a singularly perturbed system, we have the following results:

1. *The $(k-1)$-dimensional set $W^s_{loc}(M_f)$ of singular canards perturbs to a $(k-1)$-dimensional set of maximal canards for sufficiently small $\varepsilon \ll 1$.*
2. *For $-1/\mu \notin \{1\} \cup \{2m\}_{m \in \mathbb{N}}$, the $(k-1)$-dimensional set $W^u_{loc}(M_f)$ of singular faux canards perturbs to a $(k-1)$-dimensional set of maximal faux canards called* primary faux canards *for sufficiently small $\varepsilon \ll 1$.*

For $\mu < -1$, there is a one-to-one correspondence between singular and maximal canards (vrai and faux). These maximal canards form separatrix sets for solutions on $S_{a,\varepsilon}$ that either reach the vicinity of the set F locally near the set M_f (where they are able to switch from slow to fast dynamics) or not. Thus they play an important role in the analysis of excitability; see Sect. 6.4 and [80, 81, 113] for further details.

It was shown in [80], $k = 2$, that the dynamics near a pseudo saddle singularity for $-1 < \mu < 0$ become very intricate due to the existence of additional secondary faux canards.

Theorem 6.4 (cf. [80])

In the pseudo saddle case of a singularly perturbed system, if $\mu \in (-1, 0) \setminus \bigcup_{j \in \mathbb{N}_0} U_j$, where $\bigcup_{j \in \mathbb{N}_0} U_j$ is a union of sufficiently small disjoint open intervals with $-1 \in U_0$ and $-\dfrac{1}{2j} \in U_j$, $j \in \mathbb{N}$, then the singularly perturbed system possesses additional

$$n_\alpha = 2 \max\{0, \lfloor -\tfrac{1}{2\mu} \rfloor - 1\} + 1$$

$(k-1)$-dimensional sets of secondary faux canards (α-type).

For $-1 < \mu < 0$, these $(k-1)$-dimensional sets of secondary α-faux canards create some counter-intuitive geometric properties of the invariant manifolds $S_{r/s,\varepsilon}$ and $S_{a,\varepsilon}$ near the set of pseudo saddle singularities M_f. In particular, the $(k-1)$-dimensional set of primary faux canards forms locally an 'axis of rotation' for the k-dimensional sets $S_{r/s,\varepsilon}$ and $S_{a,\varepsilon}$ and hence also for the set of secondary α-faux canards; this follows from [80], case $k = 2$. These rotations happen in an $O(\sqrt{\varepsilon})$ neighbourhood of M_f. The rotational properties of solutions near faux canards are summarised in the following result:

Proposition 6.1 (cf. [80])

For $-1 < \mu < 0$, solutions in the vicinity of the pseudo saddle singularity possess no rotations about the maximal canard, but up to $\lfloor \frac{\mu-1}{2\mu} \rfloor$ rotations about the primary faux canard.

Remark 6.17 *In the case* $-1 < \mu < 0$, α-*faux canards are only the beginning of the story regarding the observed complex behaviour near a pseudo saddle. There are additional* β-*faux canards and the so-called switching or* χ-*solutions, i.e., solutions which follow close to a maximal canard but then switch to a (primary or secondary) faux canard, that provide an even richer source of rotational possibilities near a pseudo saddle singularity than in the pseudo node case (see next section). This goes beyond the scope of this manuscript and we refer the reader to [80] for further details.*

6.3.2 Pseudo Nodes

In the real eigenvalue ratio case $\mu > 0$, the set of pseudo singularities M_f is locally a $(k-2)$-dimensional normally hyperbolic manifold of pseudo nodes. We assume that the nontrivial eigenvalues of this set of pseudo nodes are negative. Thus the k-dimensional phase space S of the corresponding desingularised problem (4.20) is locally foliated by the two-dimensional stable fibres of M_f.

Definition 6.8 *The local stable manifold of a normally hyperbolic manifold of pseudo nodes* M_f, *denoted by* $W^s_{loc}(M_f)$, *is the (disjoint) union*

$$W^s_{loc}(M_f) = \bigcup_{z_{ps} \in M_f} W^s_{loc}(z_{ps}), \qquad (6.40)$$

where $W^s_{loc}(z_{ps})$ *is the local stable manifold of the base point* $z_{ps} \in M_f$ *of the desingularised problem (4.20). These sets* $W^s_{loc}(z_{ps})$ *form a family of locally invariant nonlinear fibres for* $W^s_{loc}(M_f)$ *that are tangent to the span of the two stable eigendirections of the pseudo singularity* $z_{ps} \in M_f$.

Definition 6.9 *The set* $W^{ss}_{loc}(M_f) \subset W^s_{loc}(M_f)$ *denotes the* $(k-1)$-*dimensional unique fast nonlinear sub-foliation contained in the* k-*dimensional stable foliation* $W^s_{loc}(M_f)$. *The corresponding strong stable fibres are tangent to the span of the strong stable eigendirection at each base point* $z_{ps} \in M_f$, *i.e.,*

$$W^{ss}_{loc}(M_f) = \bigcup_{z_{ps} \in M_f} W^{ss}_{loc}(z_{ps}), \qquad (6.41)$$

where $W^{ss}_{loc}(z_{ps})$ *denotes the corresponding nonlinear invariant manifold of the one-dimensional strong stable eigendirection of each pseudo singularity* $z_{ps} \in M_f$.

Definition 6.10 *The* $(k-1)$-*dimensional set* $W^{ss}_{loc}(M_f)$ *together with the* $(k-1)$-*dimensional set of regular jump points* F *bounds locally a* k-*dimensional sector on the stable branch* S_a, *called the singular funnel. Every trajectory*

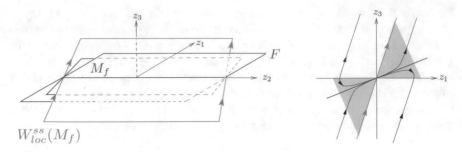

Fig. 6.13 Sketch of the local reduced flow ($k = 3$) near a local one-dimensional manifold M_f of pseudo nodes (red): (left) two-dimensional manifold of order one contact points F (blue), two-dimensional manifold $W_{loc}^{ss}(M_f)$ (green) of singular strong canards through M_f; (right) projection of reduced flow on a section $z_2 =$ constant. There exists a sector of singular canards within the singular funnel (shaded grey) bounded by $W_{loc}^{ss}(M_f)$ (green) and F (blue) which cross tangent to the weak eigendirection (brown line) of the corresponding pseudo node $z_{ps} \in M_f$

starting inside this singular funnel reaches the set of pseudo node singularities M_f in finite time and subsequently crosses the set F transversely to the 'unstable' branch $S_{r/s}$ in the direction that is tangent to the weak stable eigendirection of the corresponding pseudo node singularity $z_{ps} \in M_f$.

Thus every trajectory that originates within a singular funnel is a singular canard. Trajectories that start on the $(k-1)$-dimensional boundary set $W_{loc}^{ss}(M_f)$ reach also the set M_f in finite time but cross tangent to the strong stable eigendirection of the corresponding pseudo node singularity $z_{ps} \in M_f$.

All other trajectories of the reduced flow starting on S (close to F) outside the singular funnel reach the set of regular jump points $F \backslash M_f$ in finite forward (or backward) time where they cease to exist due to finite time blow-up.

Remark 6.18 *In the case $k = 2$, we deal with isolated pseudo nodes. The one-dimensional strong canard $W_{loc}^{ss}(z_{ps})$ of a pseudo node $z_{ps} \in M_f$ together with the one-dimensional manifold of contact points F form the boundary of a singular funnel. All solutions within the singular funnel are singular canards that are able to cross with finite speed from one branch of S to another via the isolated pseudo node z_{ps}; compare with Fig. 6.13 (right).*

Remark 6.19 *A singular funnel of faux canards can be defined in the obvious way.*

Again, all standard results on pseudo (folded) node singularities and associated canards apply via local equivalence to the general case.

Theorem 6.5 (cf. [14, 102, 110, 111]) *In the pseudo node case $0 < \mu \leq 1$ of a singularly perturbed system we have the following results:*

(i) The $(k-1)$-dimensional set $W_{loc}^{ss}(M_f)$ of singular strong canards perturbs to a $(k-1)$-dimensional set of maximal strong canards called the set of primary strong canards for sufficiently small $\varepsilon \ll 1$.

(ii) If $1/\mu \notin \mathbb{N}$ then the $(k-1)$-dimensional set of singular weak canards perturbs to a $(k-1)$-dimensional set of maximal weak canards called the set of primary weak canards for sufficiently small $\varepsilon \ll 1$.

(iii) If $2m+1 < \mu^{-1} < 2m+3$, $m \in \mathbb{N}$ and $\mu^{-1} \neq 2m+2$, then there exist m additional sets of maximal canards, all $(k-1)$-dimensional, called the sets of secondary canards for sufficiently small $\varepsilon \ll 1$. These m sets of secondary canards are $O(\varepsilon^{(1-\mu)/2})$ close to the set of primary strong canards in an $O(1)$ distance from the set F.

In the pseudo node case, only a finite number of sets of maximal canards persist under small perturbations $0 < \varepsilon \ll 1$ out of the continuum of singular canards within the singular funnel. These sets of maximal canards create some counter-intuitive geometric properties of the invariant manifolds $S_{a,\varepsilon}$ and $S_{r/s,\varepsilon}$ near the set of pseudo singularities M_f. In particular, the $(k-1)$-dimensional set of primary weak canards forms locally an 'axis of rotation' for the k-dimensional sets $S_{a,\varepsilon}$ and $S_{r/s,\varepsilon}$ and hence also for the set of primary strong canards and the set(s) of secondary canards; this follows from [110], case $k = 2$. These rotations happen in an $O(\sqrt{\varepsilon})$ neighbourhood of F. The rotational properties of the sets of maximal canards are summarised in the following result:

Theorem 6.6 (cf. [14, 102, 110, 111]) In the pseudo node case of a singularly perturbed system with $2m+1 < \mu^{-1} < 2m+3$, $m \in \mathbb{N}$ and $\mu^{-1} \neq 2m+2$,

(i) the set of primary strong canards twists once around the set of primary weak canards in an $O(\sqrt{\varepsilon})$ neighbourhood of F,

(ii) the j-th set of secondary canards, $1 \leq j \leq m$, twists $(2j+1)$-times around the set of primary weak canards in an $O(\sqrt{\varepsilon})$ neighbourhood of F,

where a twist corresponds to a half rotation. Thus each set of maximal canards has a distinct rotation number.

As a geometric consequence, the funnel region of the set of pseudo nodes M_f in S_a is split by the sets of secondary canards into $(m+1)$ sub-sectors I_j, $j = 1, \ldots, m+1$, with distinct rotational properties. I_1 is the sub-sector bounded by the (set of) primary strong canard(s) and the first (set of) secondary canard(s), I_2 is the sub-sector bounded by the first and second (sets of) secondary canard(s), I_l is the sub-sector bounded by the $(m-1)$-th and the m-th (set of) secondary canard(s) and finally, I_{m+1} is bounded by the m-th (set of) secondary canard(s) and the set F. Trajectories with initial conditions in the interior of I_j, $1 \leq j < m+1$, make $(2j + 1/2)$ twists around the (set of) primary weak canard(s), while trajectories with initial conditions in the interior of I_{m+1} make at least $[2(m + 1) - 1/2]$ twists around the (set of) primary weak canard(s). All these solutions are forced to follow the *funnel*

created by the manifolds $S_{a,\sqrt{\varepsilon}}$ and $S_{r/s,\sqrt{\varepsilon}}$. After solutions leave the funnel
in an $O(\sqrt{\varepsilon})$ neighbourhood of F they get repelled by the manifold $S_{r/s,\sqrt{\varepsilon}}$
and will follow close to the set of fast fibres. Hence, pseudo node type canards
form separatrix sets in the phase space for different rotational properties near
critical manifolds with contact structure.

Remark 6.20 *Canard induced mixed-mode oscillations (MMOs) are a prominent example of a complex rhythm that can be traced to pseudo node singularities. A three-dimensional extension of the autocatalator model (2.54) by
Petrov et al. [90] is known as the prototypical chemical model for mixed-mode
oscillatory behaviour. Milik and Szmolyan [77] were the first who studied this
model by means of geometric singular perturbation theory. We refer the interested reader to a MMO review [25] and the references therein for the vast
literature on this subject matter.*

6.3.3 Pseudo Foci

In the case $\mu \in \mathbb{C}\backslash\mathbb{R}$, all solutions of the reduced problem starting on S
(close to F) reach the set of regular jump points $F\backslash M_f$ in finite forward or
backward time where they cease to exist due to finite time blow-up.

Remark 6.21 *There exist no singular canards near a pseudo focus. Hence
there are no canards in this case.*

6.3.4 Pseudo Saddle-Nodes

Pseudo saddles, pseudo nodes and pseudo foci are not related to equilibria
of the reduced problem. Those pseudo singularities that are related to equilibria of the reduced problem (and, hence, to equilibria of the full system)
are necessarily codimension-one and form a subset of the codimension-one
bifurcation structure in the desingularised problem associated with the case
$\mu = 0$, where one of the nontrivial eigenvalues vanishes. They are called
pseudo saddle-nodes (PSN) and split into three subtypes: type I, II and III.
In the standard case, they are known as folded saddle-nodes.

Only two of these subtypes are related to equilibria: PSN II and PSN
III. A pseudo (folded) saddle-node *type II* [58] indicates a transcritical bifurcation of a pseudo (folded) singularity and an equilibrium, while a pseudo
(folded) saddle-node *type III* [94] indicates a pitchfork bifurcation of two
pseudo (folded) singularities and an equilibrium.

The pseudo (folded) saddle-node *type I* [108] is a saddle-node bifurcation
of two pseudo (folded) singularities, i.e., no equilibria are involved.

6.4 Two-Stroke Excitability with Dynamic Parameter Variation

Recall the extended autonomous system (6.36) of the two-stroke oscillator model with dynamic parameter variation which is of the general form (3.10) with $N(s, x, y)$ and $f(s, x, y)$ given in (6.37), and $G(s, x, y)$ in (6.38). In the following we perform a geometric singular perturbation analysis of this system to explain the observed excitability dynamics shown in Fig. 6.11.

Layer Problem:
The set of singularities, $N(s, x, y) f(s, x, y) = 0$, is given by

$$S_0 = S \cup L = \{y(s, x) = \delta(s)\} \cup \{x(s) = y(s) = 0\}.$$

Note, s represents a parameter in the layer problem, i.e., it is considered a slow variable of the extended system (6.36). Thus the set S_0 contains a (standard) one-dimensional critical manifold

$$L = \{(s, x, y) \in \mathbb{R}^3 \ : \ x(s) = y(s) = 0\}, \tag{6.42}$$

which is a normally hyperbolic repelling one-dimensional manifold since the two nontrivial eigenvalues of the Jacobian $Dh|_L$ are complex conjugates with positive real part.[10]
The set S_0 also contains a subset S which forms a two-dimensional critical manifold

$$S = \{(s, x, y) \in \mathbb{R}^3 \ : \ y(s, x) = \delta(s)\} \tag{6.43}$$

of this problem. The Jacobian Dh evaluated along $(s, x, y) \in S$ is given by

$$Dh|_S = \begin{pmatrix} 0 & 0 & 0 \\ \delta(s)\delta'(s) & 0 & -\delta(s) \\ (x - \alpha\delta(s))\delta'(s) & 0 & x - \alpha\delta(s), \end{pmatrix}$$

which has two trivial eigenvalues and $\lambda_1 = Df N|_S = x - \alpha\delta(s)$, $\forall (s, x) \in \mathbb{R}^2$, as its nontrivial eigenvalue. Thus S loses normal hyperbolicity for $x = \alpha\delta(s)$, and $S = S_a \cup F \cup S_r$ consists of an attracting branch S_a for $x < \alpha\delta(s)$, a repelling branch for $x > \alpha\delta(s)$ and a one-dimensional manifold of contact points

$$F = \{(s, x, y) \in S \ : \ x = \alpha\delta(s)\}. \tag{6.44}$$

The conditions (4.6) of Lemma 4.1 are fulfilled, i.e.,

---

[10] While the set L does not constitute a set of isolated singularities of $N(s, x, y)$ as stated in Assumption 3.3, the theory presented is still valid since s is a 'true' slow variable.

$$\operatorname{rk} Df|_F = n - k = 1\,,$$

$$l \cdot (D^2 f(Nr, Nr) + Df\,DN(Nr, r)) = D^2 f(N, N)|_F + Df\,DNN|_F$$

$$= 0 + \begin{pmatrix} \delta'(s) & 0 & -1 \end{pmatrix} \begin{pmatrix} 0 & 0 & 0 \\ 0 & 0 & 1 \\ 0 & -1 & \alpha \end{pmatrix} \begin{pmatrix} 0 \\ \delta \\ 0 \end{pmatrix} = \delta \neq 0\,,$$

which shows that the contact between the layer flow and the critical manifold S along F is of order one. Note that the fast rotations in (x, y)-space around the line L cause the order one contact of the layer flow with S along F. Note further that $S \cap L = \emptyset$ by Model Assumption 6.2; see Fig. 6.14 for the corresponding geometric configuration.

Remark 6.22 *The temporal evolution of the dynamic parameter variation* $\delta(s)$ *is encoded in the geometry of the two-dimensional critical manifold S, i.e., the ramp is now a geometric feature of S; see Fig. 6.14.*

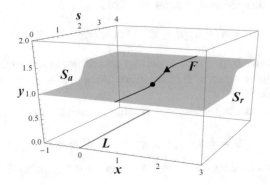

Fig. 6.14 $\alpha = 1.0$, $\beta = 0.5$, $\gamma = 1.0$: critical manifold S (6.43), grey, with contact manifold F (6.44), red, and isolated line of singularities L (6.42), orange. F is defined by the red ramp from Fig. 6.11

Reduced Problem:
The reduced flow along the one-dimensional normally hyperbolic repelling line L (6.42) is simply given by $\dot{s} = 1$, i.e., reflecting the evolution of slow time.

For the reduced flow along the two-dimensional manifold S (6.43), we study the corresponding desingularised problem (4.20) which is given by

$$\begin{pmatrix} \dot{s} \\ \dot{x} \\ \dot{y} \end{pmatrix} = \left(-(x - \alpha\delta(s)) \begin{pmatrix} 1 & 0 & 0 \\ 0 & 1 & 0 \\ 0 & 0 & 1 \end{pmatrix} + \begin{pmatrix} 0 & 0 & 0 \\ \delta(s)\delta'(s) & 0 & -\delta(s) \\ (x - \alpha\delta(s))\delta'(s) & 0 & x - \alpha\delta(s) \end{pmatrix} \right) \begin{pmatrix} 1 \\ 0 \\ -\beta + \gamma x \end{pmatrix}$$

evaluated along the critical manifold S (6.43) given as a graph over the (s, x)-coordinate chart. We study the desingularised flow in this single coordinate chart:

$$\dot{s} = \alpha\delta(s) - x$$
$$\dot{x} = \delta(s)(\beta - \gamma x + \delta'(s)). \tag{6.45}$$

Since $\lambda_1 = \det(DfN)|_S = x - \alpha\delta(s)$, the flow of this desingularised system has to be reversed on S_r, i.e., for $x > \alpha\delta(s)$, to obtain the corresponding reduced flow. By Definition 6.2, the set of (isolated) pseudo singularities is given by

$$M_f = \{(s, x, y) \in F : \delta'(s) - \alpha\gamma\delta(s) + \beta = 0\}. \tag{6.46}$$

The type of these pseudo singularities $z_{ps} \in M_f$ is obtained by calculating the corresponding Jacobian of system (6.45),

$$J = \begin{pmatrix} \alpha\delta' & -1 \\ \delta'(\beta - \gamma x + \delta') + \delta\delta'' & -\gamma\delta \end{pmatrix}\Bigg|_{z_{ps} \in M_f} = \begin{pmatrix} \alpha\delta' & -1 \\ \delta\delta'' & -\gamma\delta \end{pmatrix}. \tag{6.47}$$

The trace and the determinant of the Jacobian (6.47) evaluated along a pseudo singularity are given by

$$\mathrm{tr}(J) = \alpha\delta' - \gamma\delta \quad \text{and} \quad \det(J) = \delta(\delta'' - \alpha\gamma\delta'). \tag{6.48}$$

Depending on the specific choice of the slowly varying parameter $\delta(s)$ that fulfil the Model Assumption 6.2, we are able to find all types of pseudo singularities. In the case of the ramp (6.34), we have $\delta(s) > 0$ and $\delta'(s) \geq 0$ for all $s \in \mathbb{R}$ while the curvature $\delta''(s)$ is changing its sign (from positive to negative) due to a single inflection point in the ramp. If, e.g., a pseudo saddle singularity exists, then it must be located on the contact manifold F (6.44) after the inflection point of F (which is the same as for the ramp) since this is a necessary condition for $\det(J) < 0$ in (6.48). Similarly, if, e.g., a pseudo focus singularity exists, then it must be located on the contact manifold F (6.44) before the inflection point of F since this is a necessary condition for $\mathrm{tr}^2(J) - 4\det(J) < 0$ in (6.48).

For the two ramps used in Fig. 6.11, the corresponding reduced flow on S projected onto the (s, x)-coordinate chart is shown in Fig. 6.15. Along the one-dimensional contact manifold F, we find two pseudo singularities: a pseudo focus located before the inflection point of F and a pseudo saddle located after the inflection point of F, as predicted. Note, $\dot{x} > 0$ only along the segment of F bounded by the two pseudo singularities. To reach this segment of F and, hence, to switch from slow to fast dynamics and to be able to elicit a spike, the initial state (s^-, x_{rest}^-) must be in the 'domain of attraction' of this segment. This domain is bounded by the sector formed by the pseudo saddle canard (blue) on S_a and the segment of F to the left of the pseudo saddle. Thus the pseudo saddle canard forms a threshold manifold on S_a. The initial state on the attracting branch S_a of the critical manifold is (s^-, x_{rest}^-), where $x = x_{rest}^-$ corresponds to the original excitable rest state of system (6.30) with $\delta = \delta_-$ fixed which is $x_{rest}^- = \beta/\gamma = 0.5$ by (6.31) for our specific example.

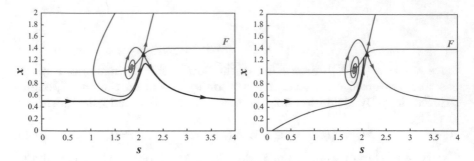

Fig. 6.15 Reduced flow shown projected onto coordinate chart (s, x) corresponding to the ramp protocols shown in Fig. 6.11: there are two pseudo singularities along the curve F, a pseudo saddle (black triangle) and a pseudo focus (grey circle), the pseudo saddle canard (blue) that crosses from the lower stable branch S_a via the pseudo saddle singularity onto the repelling upper branch S_r. This canard forms the firing threshold manifold. Note, the other canard of the pseudo saddle singularity (red) crosses from S_r to S_a—it is a faux canard. The segment of the faux canard on S_a forms a boundary that prevents trajectories to the right of the blue canard to spike. For the same initial condition (the rest state), the canards explain why there are two significant different responses: (left) none as shown in Fig. 6.11, blue ramp/trajectory; (right) spike as shown in Fig. 6.11, red ramp/trajectory, since the trajectory reaches F where it jumps off S

As can be also seen in Fig. 6.15, the position of the canard changes as the (maximal) slope of $\delta(s)$ changes. Clearly, pseudo singularities and their canards encode the complete temporal information of $\delta(s)$, i.e., amplitude, slope, curvature, etc. The configuration in Fig. 6.15 (left) predicts no spike while Fig. 6.15 (right) predicts a spike. These correspond to the two cases of the ramp protocols shown in Fig. 6.11.

Theorem 6.3 guarantees the existence of a maximal canard of pseudo saddle type for $0 < \varepsilon \ll 1$ which forms the excitability separatrix in the full problem. It is remarkable that dynamic, nonautonomous information such as the evolution profile of the slowly varying parameter $\delta(s)$ is encoded in the location of an invariant manifold of a singular perturbation problem—a pseudo saddle canard. Here, in particular, we observe that only changing the slope is sufficient to elicit a spike. Thus our analysis provides a *slow modulation* condition for transient phenomena based on canard theory.

Remark 6.23 *For the specific ramp (6.34), increasing the parameter s_1 decreases the steepness of the ramp (while the height stays constant). In the corresponding reduced problem, we observe two pseudo singularities (see Fig. 6.15) that approach each other as s_1 increases until they annihilate each other for $s_1^{sn1} \approx 0.584$ in a pseudo saddle-node type I bifurcation. This implies that the pseudo focus shown in Fig. 6.15 has to change into a pseudo node before this PSN I bifurcation can happen. Interesting local dynamics are expected close to PSN I, i.e., $s_1 < s_1^{sn1}$ where a pseudo saddle and a pseudo node exist, based on the results presented in Sects. 6.3.1 and 6.3.2 on local* *oscillatory behaviour near pseudo saddles and nodes. We refer the interested reader to [79, 108].*

Remark 6.24 *The existence of a pseudo singularity is a necessary but not a sufficient condition for an excitable system to elicit a spike under slow dynamic changes of a system parameter. At the heart of the issue lies the relative position of the pseudo saddle canard that forms the threshold manifold in these models relative to the equilibrium state of the underlying excitable system before the slow modulation kicks in (which is the initial condition when the modulation is turned on).*

Dynamic forcing has the potential to create pseudo singularities and to form these effective separatrices or to change the global return mechanism. Hence, the specific nature of the dynamic forcing determines which local attractor states can be reached through global mechanisms. This point of view has profound consequences in the analysis of excitable systems. Physiological problems are prominent examples where this has been observed such as in, e.g., auditory brain stem neurons [75], modelling propofol anaesthesia [81] and cell calcium dynamics [37]. Another striking example comes from climate science and the study of the compost bomb instability by Wieczorek et al. [115]. For a review on excitability and GSPT (in standard form) see also [113].

Chapter 7
What We Did Not Discuss

Finally, we briefly mention a few selected topics on GSPT that have not been covered in this manuscript. This list of topics is non-inclusive—it is an author's choice (like all topics covered in this manuscript).

7.1 The 'Other' Delayed Loss of Stability Phenomenon

Pseudo singularities and associated canards discussed in Chap. 6 are one class of examples of the *delayed loss of stability* phenomenon where trajectories of a singularly perturbed system cross from an attracting branch S_a to a 'repelling' branch $S_{r/s}$ and follow it for an $O(1)$ slow time amount before the delayed repulsion kicks in.

As mentioned at the beginning of Chap. 4, loss of normal hyperbolicity of the critical manifold S is not only related to a real nontrivial eigenvalue crossing zero—the set $F = S \backslash S_n$—but also to a pair of complex-conjugate nontrivial eigenvalues crossing the imaginary axis—the set $AH = S_n \backslash S_h$. Note, the determinant of the Jacobian $DfN|_S$ does *not* change sign near this subset $AH \subset S_n$. Hence, the splitting $T_z\mathbb{R}^n = T_z S \oplus N_z$, $\forall z \in S_n$, is preserved near AH, and so is the associated (unique) projection operator Π^S. Consequently, the reduced flow near AH is described by Lemma 3.4 as for the normally hyperbolic case.

More importantly, the subset $AH \subset S_n$ does not form an 'obstacle' for the reduced flow to cross (compared to the subset $F = S \backslash S_n$), i.e., slow motion from an attracting branch S_a to a 'repelling' branch $S_{r/s}$ via any point $z \in AH$ (or vice versa) is generically possible. Thus in this case of loss of normal hyperbolicity, *singular canard-type* behaviour is the norm; compared with the Definition 6.2 in the case of pseudo singularities $z_{ps} \in M_F \subseteq F$.

© Springer Nature Switzerland AG 2020
M. Wechselberger, *Geometric Singular Perturbation Theory Beyond the Standard Form*, Frontiers in Applied Dynamical Systems: Reviews and Tutorials 4, https://doi.org/10.1007/978-3-030-36399-4_7

On the other hand, there exist no associated *maximal canards* in the AH case, i.e., there is no intersection of the invariant manifolds $S_{a,\varepsilon}$ and $S_{r/s,\varepsilon}$ near AH for $0 < \varepsilon \ll 1$; compare with the Definition 6.4 in the case of pseudo singularities $z_{ps} \in M_f \subseteq F$ and their maximal canards. Still, a delayed loss of stability phenomenon is observed, because the manifolds $S_{a,\varepsilon}$ and $S_{r/s,\varepsilon}$ *graze* exponentially close in the neighbourhood of $AH \subset S_n$ before they reject each other at a set known as *buffer points* which is located an $O(1)$ distance away from AH; see e.g., [7] and references therein. Thus *any* solution that starts on $S_{a,\varepsilon}$ and crosses AH experiences a delayed loss of stability phenomenon. For details and results on this subject, we refer the reader to the original work by Shishkova [100] and Neishtadt [84], the conference proceedings on *dynamic bifurcations* [7] as well as to a GSPT inspired work by Hayes et al. [41].

7.2 Algorithmic and Computational Aspects of GSPT

Biochemical reaction networks are usually modelled as a large set of ordinary differential equations with an even larger set of system parameters. Based on experimental observations (or otherwise), one is often interested in quasi-steady state (QSS) phenomena with the aim to reduce the model and, hence, make it amenable to a mathematical analysis. This reduction technique is based on identifying multiple time scales involved in the reaction processes modelled, i.e., finding small dimensionless parameters related to the distinct reaction rates, as well as locating low-dimensional invariant sets within the model in an appropriate singular limit, i.e., identifying the model as a singular perturbation problem with associated critical manifold(s). Many of these models that incorporate processes evolving on disparate time scales and show multiple time-scale dynamics do not necessarily have the standard form of a singular perturbation problem, which is part of the motivation for the more general (coordinate-independent) approach to GSPT presented in this manuscript.

Many approaches in the literature require an initial assumption, or guess, about time scales involved and/or the existence of an invariant set(s). Instead of searching for small parameters in a given system that (might) lead to a singular perturbation problem (beyond the standard form), the existence of a small singular perturbation parameter is assumed. In recent work, Goeke et al. [33, 35] addressed this issue and formulated an algorithmic approach to identify small parameters and a QSS reduction in systems with a polynomial/rational vector field which is based on the normally hyperbolic theory for general geometric singular perturbation problems discussed in this manuscript.

This algorithmic approach of identifying small parameters relates directly to computational aspects for high dimensional singular perturbation problems, e.g., how to identify and calculate lower dimensional invariant manifolds. In the context of combustion modelling, Lam and Goussis [64] introduced the so-called *computational singular perturbation (CSP)* method, and Mease [74] was the first to point out its relationship to GSPT. Other well-known methods are the *zero derivative principle (ZDP)*, the *intrinsic low-dimensional manifold (ILDM)* method by Maas and Pope [73], or the method introduced by Roussel and Fraser [95]. In a series of papers, Zagaris et al. [49, 116, 117] focus on the GSPT approach to many of these methods to provide a clear mathematical interpretation of QSS as a normally hyperbolic singular perturbation phenomenon. Recently, [72] adapted the CSP method to slow-fast systems beyond the standard form.

Computational methods for singular perturbation problems where loss of normal hyperbolicity occurs such as the *canard phenomenon* are almost exclusively dealt with in standard form, see e.g., [25, 61] and references therein. There is clearly a need to extend these methods to general singular perturbation problems that show relaxation or more complex oscillatory patterns (including delayed loss of stability phenomena).

7.3 More Than Two Time Scales

For systems with multiple time scales, Cardin and Teixeira [18] provide results that extend Fenichel theory for more than two time scales. These results emphasise that the hierarchy of time scales is reflected in a hierarchy of nested invariant manifolds that are progressively approached (in the attracting case) as time progresses from one distinct scale to another $(O(1), O(\varepsilon_1), O(\varepsilon_2), \ldots, O(\varepsilon_k)$ where $\varepsilon_k \ll \ldots \ll \varepsilon_2 \ll \varepsilon_1 \ll 1)$. Thus QSS reduction techniques can be naturally extended to these multiple time-scale problems.

In the case of loss of normal hyperbolicity, the autocatalator model introduced in Sect. 2.2.4 has been identified as a three time-scale problem where the different time scales are related through a single perturbation parameter $\varepsilon \ll 1$, i.e., the model has time scales of order $O(1)$, $O(\varepsilon)$ and $O(\varepsilon^2)$. Krupa et al. [59] were the first to study such three time-scale problems in the context of mixed-mode oscillations. Other multiple time-scale models such as the glycolytic oscillator analysed by Kosiuk and Szmolyan [53], the pituitary lactotroph model by Vo et al. [109] and Letson et al. [69] or a McArthur–Rosenzweig-type model studied by Nan et al. [83] and Desroches and Kirk [24] have two distinct small parameters $\varepsilon, \delta \ll 1$, that lead to three distinct time scales in these models. Depending on specific limiting relationships between these singular perturbation parameters, certain model behaviour can be identified and analysed using GSPT techniques. These specific limits highlight the general problem of identifying small parameters in large multiple time-scale models such that model reductions and/or GSPT tools can be used to rigorously analyse such models.

7.4 Regularisation of Non-smooth Dynamical Systems

Piecewise-smooth dynamical systems [10] are commonplace in mechanics and control theory where a discontinuous change of a state variable or its derivative(s) is the norm which indicates, e.g., the crossing, sliding or grazing of a hypersurface in phase space known as a *switching manifold*. In the case where the vector field under study has a discontinuous change at such a switching manifold, the corresponding non-smooth system is usually referred to as a *Filippov system* [32]. Friction models such as stick-slip oscillators are a prime example of such non-smooth dynamical systems [43, 89].

In the last decade, there has been a significant effort to apply regularisation techniques to these non-smooth systems so that they can be analysed by the available toolkit from smooth dynamical systems theory. Such a regularisation approach leads typically to a singular perturbation problem, see e.g., [17], where different branches of locally invariant critical manifold(s) represent the 'piecewise-smooth' vector fields of the underlying non-smooth system. Bossolini et al. [11] studied, e.g., a regularised *stiction* model that takes the form of a singular perturbation problem which can be analysed using GSPT and canard theory; see also Kristiansen and Hogan [56]. The non-standard 'stick-slip' relaxation oscillator and its variation introduced in this manuscript (see Sects. 2.2.2 and 6.1.4) is another example and highlights the importance of extending GSPT beyond the standard form to understand non-smooth behaviour.

7.5 Infinite Dimensional Dynamical Systems

A final remark on the main limitation of the presented theory—its finite dimensionality. The existing GSPT literature for infinite dynamical systems is sparse. Bates et al. [5] presents a general theory for semiflows in Banach spaces, while Menon and Haller [76] focus on infinite dimensional GSPT for the Maxwell–Bloch equations. Recently, there has been work related to the delayed loss of stability phenomena in infinite dimensional dynamical systems such as neural field equation by Avitabile et al. [3] and in some spatio-temporal pattern-forming systems by Avitabile et al. [4] and by Kaper and Vo [48].

References

1. J. Anderson, A. Ferri, Behavior of a single-degree-of-freedom system with a generalized friction law. J. Sound Vib. **140**(2), 287–304 (1990)
2. J. Argemi, Approche qualitative d'un problème de perturbation singulières dans $\mathbb{R}^4$, in *Equadiff*, vol. 78 (1978), pp. 333 340
3. D. Avitabile, M. Desroches, E. Knobloch, Spatiotemporal canards in neural field equations. Phys. Rev. E **95**(4), 042205 (2017)
4. D. Avitabile, M. Desroches, E. Knobloch, M. Krupa, Ducks in space: from nonlinear absolute instability to noise-sustained structures in a pattern-forming system, in *Proceedings of the Royal Society of London A: Mathematical, Physical and Engineering Sciences*, vol. 473 (2017), p. 20170018
5. P. Bates, K. Lu, C. Zeng, Existence and persistence of invariant manifolds for semiflows in banach space. Mem. Am. Math. Soc. **645** (1998)
6. E. Benoit, Systèmes lents-rapides dans $\mathbb{R}^3$ et leur canards. Astérisque **109–110**, 159–191 (1983)
7. E. Benoit (ed.), *Dynamic Bifurcations*. Lecture Notes in Mathematics, vol. 1493 (Springer, Berlin, 1991)
8. E. Benoit, J. Callot, F. Diener, M. Diener, Chasse aux canards. Collect. Math. **31–32**, 37–119 (1981)
9. E. Berger, Friction modelling for dynamic system simulation. Appl. Mech. Rev. **55**(6), 535–577 (2002)
10. M. Bernardo, C. Budd, A. Champneys, P. Kowalczyk, Piccewise-smooth dynamical systems: theory and applications, in *Applied Mathmematical Sciences* (Springer, London, 2008)
11. E. Bossolini, M. Brøns, K. Kristiansen, Canards in stiction: on solutions of a friction oscillator by regularization. SIAM J. Appl. Dyn. Syst. **16**(4), 2233–2256 (2017)
12. H. Broer, T. Kaper, M. Krupa, Geometric desingularization of a cusp singularity in slow-fast systems with applications to Zeeman's examples. J. Dyn. Diff. Equat. **25**, 925–958 (2013)

M. Wechselberger, *Geometric Singular Perturbation Theory Beyond the Standard Form*, Frontiers in Applied Dynamical Systems: Reviews and Tutorials 4, https://doi.org/10.1007/978-3-030-36399-4

13. M. Brøns, Canard explosion of limit cycles in templator models of self-replication mechanisms. J. Chem. Phys. **134**(14), 144105 (2011)

14. M. Brøns, M. Krupa, M. Wechselberger, Mixed mode oscillations due to the generalized canard phenomenon, in *Fields Institute Communications*, vol. 49, pp. 39–63 (Fields Institute, Ontario, 2006)

15. M. Brøns, T. Kaper, H. Rotstein, *Introduction to Focus Issue: Mixed Mode Oscillations: Experiment, Computation, and Analysis* (Chaos, Woodbury, 2008), p. 015101

16. A.D. Bruno, *Local Methods in Nonlinear Differential Equations* (Springer, Berlin, 1989)

17. P. Cardin, P. da Silva, M. Teixeira, On singularly perturbed Filippov systems. Eur. J. Appl. Math. **24**, 835–856 (2013)

18. P. Cardin, M. Teixeira, Fenichel theory for multiple time scale singular perturbation problems. SIAM J. Appl. Dyn. Syst. **16**, 1425–1452 (2017)

19. J. Carr, *Applications of Centre Manifold Theory* (Springer, Berlin, 1981)

20. R. Courant, D. Hilbert, *Methods of Mathematical Physics* (Interscience, New York, 1964)

21. P. De Maesschalck, F. Dumortier, Slow-fast Bogdanov-Takens bifurcations. J. Differ. Equ. **250**, 1000–1025 (2011)

22. P. De Maesschalck, F. Dumortier, M. Wechselberger, Special issue on bifurcation delay. DCDS-S **2**(4), 723–1023 (2009)

23. P. De Maesschalck, F. Dumortier, R. Roussarie, Cyclicity of common slow–fast cycles. Indag. Math. **22**, 165–206 (2011)

24. M. Desroches, V. Kirk, Spike-adding in a canonical three time scale model: superslow explosion and folded-saddle canards. SIAM J. Appl. Dyn. Syst. **17**, 1989–2017 (2018)

25. M. Desroches, J. Guckenheimer, B. Krauskopf, C. Kuehn, H. Osinga, M. Wechselberger, Mixed-mode oscillations with multiple time scales. SIAM Rev. **54**(2), 211–288 (2012)

26. J. Dietrich, M. Linker, Fault stability under conditions of variable normal stress. Geophys. Res. Lett. **19**, 1691–1694 (1992)

27. F. Dumortier, Structures in dynamics: finite dimensional deterministic systems, chap, in *Local Study of Planar Vector Fields: Singularities and their Unfoldings* (Studies in Mathematical Physics, North Holland, 1991)

28. F. Dumortier, R. Roussarie, Canard cycles and center manifolds. Mem. Am. Math. Soc. **121**(577), x+100 (1996)

29. T. Erneux, A. Goldbeter, Rescue of the quasi-steady state approximation in a model of enzymatic cascade. SIAM J. Appl. Math. **67**, 305–320 (2006)

30. N. Fenichel, Geometric singular perturbation theory for ordinary differential equations. J. Differ. Equ. **31**(1), 53–98 (1979)

31. B. Fiedler, S. Liebscher, Global analysis of dynamical systems, chap, in *Takens-Bogdanov Bifurcations Without Parameters and Oscilatory Shock Profiles* (Institute of Physics, Bristol, 2001), pp. 211–260

32. A. Filippov, Differential equations with discontinuous righthand sides, in *Mathematics and its Applications* (Kluwer, Dordrecht, 1988)

33. A. Goeke, S. Walcher, Quasi steady-state: searching for and utilizing small parameters, in *Springer Proceedings in Mathematics and Statistics of Recent Trends in Dynamical Systems*, vol. 35 (eds.) by A. Johann et al. (Springer, Berlin, 2013), pp. 153–178

34. A. Goeke, S. Walcher, A constructive approach to quasi-staedy state reductions. J. Math. Chem. **52**, 2596 2626 (2014)

35. A. Goeke, S. Walcher, E. Zerz, Determining 'small parameters' for quasi-steady state. J. Differ. Equ. **259**, 1149–1180 (2015)

36. I. Gucwa, P. Szmolyan, Geometric singular perturbation analysis of an autocatalator model. Discrete Contin. Dynam. Syst. Ser. S **2**(4), 783–806 (2009)

37. E. Harvey, V. Kirk, H. Osinga, J. Sneyd, M. Wechselberger, Understanding anomalous delays in a model of intracellular calcium dynamics. Chaos **20**, 045104 (2010)

38. K. Harley, P. van Heijster, R. Marangell, G. Pettet, M. Wechselberger, Existence of travelling wave solutions for a model of tumour invasion. SIAM J. Appl. Dyn. Syst. **13**, 366–396 (2014)

39. K. Harley, P. van Heijster, R. Marangell, G. Pettet, M. Wechselberger, Novel solutions for a model of wound healing angiogenesis. Nonlinearity **27**, 2975–3003 (2014)

40. K. Harley, P. van Heijster, R. Marangell, G. Pettet, M. Wechselberger, Numerical computation of an Evans function for travelling waves. Math. Biosci. **266**, 36–51 (2015)

41. M. Hayes, T. Kaper, P. Szmolyan, M. Wechselberger, Geometric desingularization of degenerate singularities in the presence of rotation: a new proof of known results for slow passage through Hopf bifurcation. Indag. Math. **27**, 1184–1203 (2016)

42. C. He, S. Ma, J. Huang, Transition between stable sliding and stick-slip due to variation in slip rate under variable normal stress condition. Geophys. Res. Lett. **25**(17), 3235–3238 (1998)

43. N. Hinrichs, M. Oestreich, K. Popp, On the modelling of friction oscillators. J. Sound Vib. **216**, 435–459 (1998)

44. O. Ilina, P. Friedl, Mechanisms of collective cell migration at a glance. J. Cell Sci. **122**, 3203–3208 (2009)

45. S. Jelbart, M. Wechselberger, Two-stroke relaxation oscillations (2019). arXiv:1905.06539

46. C.K.R.T. Jones, Geometric singular perturbation theory, in *Dynamical Systems (Montecatini Terme, 1994)*. Lecture Notes in Mathematics, vol. 1609 (Springer, Berlin, 1995), pp. 44–118

47. P.I. Kaleda, Singular systems on the plane and in space. J. Math. Sci.
    179(4), 475–490 (2011)
48. T. Kaper, T. Vo, Delayed loss of stability due to the slow passage
    through Hopf bifurcation in reaction–diffusion equations. Chaos **28**,
    091103 (2018)
49. H. Kaper, T. Kaper, A. Zagaris, Geometry of the computational sin-
    gular perturbation method. Math. Model. Nat. Phenom. **10**(3), 16–30
    (2015)
50. J. Keener, J. Sneyd, *Mathematical Physiology*, 2nd edn. (Springer,
    Berlin, 2008)
51. E.F. Keller, L.A. Segel, Model for chemotaxis. J. Theoret. Biol. **30**,
    225–234 (1971)
52. I. Kosiuk, *Relaxation Oscillations in Slow-Fast Systems Beyond the
    Standard Form.* Ph.D. thesis (University of Leipzig, Leipzig, 2012)
53. I. Kosiuk, P. Szmolyan, Scaling in singular perturbation problems:
    blowing up a relaxation oscillator. SIAM J. Appl. Dyn. Syst. **10**(4),
    1307–1343 (2011)
54. I. Kosiuk, P. Szmolyan, Geometric analysis of the Goldbeter minimal
    model for the embryonic cell cycle. J. Math. Biol., **72**(5), 1337–1368
    (2015)
55. K. Kristiansen, Blowup for flat slow manifolds. Nonlinearity **30**, 2138–
    2184 (2017)
56. K. Kristiansen, S. Hogan, On the use of blowup to study regularizations
    of singularities of piecewise smooth dynamical systems in $\mathbb{R}^3$. SIAM J.
    Appl. Dyn. Syst. **14**, 382–422 (2015)
57. M. Krupa, P. Szmolyan, Relaxation oscillation and canard explosion.
    J. Differ. Equ. **174**(2), 312–368 (2001)
58. M. Krupa, M. Wechselberger, Local analysis near a folded saddle-node
    singularity. J. Differ. Equ. **248**, 2841–2888 (2010)
59. M. Krupa, N. Popović, N. Kopell, Mixed-mode oscillations in three
    time-scale systems: a prototypical example. SIAM J. Appl. Dyn. Syst.
    7(2), 361–420 (2008)
60. C. Kuehn, Normal hyperbolicity and unbounded critical manifolds.
    Nonlinearity **27**, 1351 (2014)
61. C. Kuehn, *Multiple Time Scale Dynamical Systems* (Springer, Berlin,
    2015)
62. C. Kuehn, P. Szmolyan, Multiscale geometry of the Olsen model and
    non-classical relaxation oscillations. J. Nonlinear Sci. **25**(3), 583–629
    (2015)
63. Y. Kuznetsov, *Elements of Applied Bifurcation Theory*, vol. 112
    (Springer, New York, 1995)
64. S. Lam, D. Goussis, Understanding complex chemical kinetics with
    computational singular perturbation, in *Proceedings of the Twenty-
    Second Symposium (International) on Combustion*, pp. 931–941 (1988)

65. K. Landman, M. Simpson, G. Pettet, Tactically-driven nonmonotone travelling waves. Physica D **237**, 678–691 (2008)
66. P. Lax, *Hyperbolic Partial Differential Equations* (American Mathematical Society/Courant Institute of Mathematical Sciences, New York, 2006)
67. P. Le Corbeiller, Two-stroke oscillator, in *IRE Transactions on Curcuit Theory* (1960), pp. 387–398
68. J. Lee, *Introduction to Smooth Manifolds*, 2nd edn. Graduate Texts in Mathematics (Springer, New York, 2013)
69. B. Letson, J. Rubin, T. Vo, Analysis of interacting local oscillation mechanisms in three-timescale systems. SIAM J. Appl. Math. **77**, 1020–1046 (2017)
70. S. Liebscher, *Bifurcation without Parameters* (Springer, Berlin, 2015)
71. C.C. Lin, L.A. Segel, *Mathematics Applied to Deterministic Problems in the Natural Sciences* (SIAM, Philadelphia, 1988)
72. I. Lizarraga, M. Wechselberger, Computational singular perturbation method for nonstandard slow-fast systems (2019). arXiv:1906.06049
73. U. Maas, S. Pope, Simplifying chemical kinetics: intrinsic low-dimensional manifolds in composition space. Combust. Flame **88**, 239–264 (1992)
74. K. Measc, Geometry of computational singular perturbations, in *IFAC Proceedings*, vol. 28 (1995), pp. 855–861
75. X. Meng, G. Huguet, J. Rinzel, Type III excitability, slope sensitivity and coincidence detection. Discrete Contin. Dynam. Syst. **32**(8), 2729–2757 (2012)
76. G. Menon, G. Haller, Infinite dimensional geometric singular perturbation theory for the Maxwell–Bloch equations. SIAM J. Math. Anal. **33**, 315–346 (2001)
77. A. Milik, P. Szmolyan, Multiple time scales and canards in a chemical oscillator, in *Multiple-Time-Scale Dynamical Systems*, vol. 122 (IMA, Minneapolis, 1997), pp. 117–140
78. E. Mishchenko, Y. Kolessov, A. Kolessov, N. Rozov, *Asymptotic Methods in Singularly Perturbed Systems* (Consultants Bureau, New York, 1994)
79. J. Mitry, *A Geometric Singular Perturbation Approach to Neural Excitability: Canards and Firing Threshold Manifolds* (The University of Sydney, Camperdown, 2016). Ph.D. thesis
80. J. Mitry, M. Wechselberger, Folded saddles and faux canards. SIAM J. Appl. Dyn. Syst. **16**, 546–596 (2017)
81. J. Mitry, M. McCarthy, N. Kopell, M. Wechselberger, Excitable neurons, firing threshold manifolds and canards. J. Math. Neurosci., **3**(1), 12 (2013)
82. J. Murray, *Mathematical Biology* (Springer, Berlin, 2002)

83. P. Nan, Y. Wang, V. Kirk, J. Rubin, Understanding and distinguishing three time scale oscillations: case study in a coupled Morris-Lecar system. SIAM J. Appl. Dyn. Syst. **14**, 1518–1557 (2015)

84. A.I. Neishtadt, Persistence of stability loss for dynamical bifurcations. Differ.Uravn. **23**(12), 2060–2067 (1987)

85. L. Noethen, S. Walcher, Tikhonov's theorem and quasi-steady state. Discrete Contin. Dynam. Syst. Ser. B **16**(3), 945–961 (2011)

86. B. Novak, J. Tyson, Design principles of biochemical oscillators. Nat. Rev. Mol. Cell Biol. **9**(12), 981–991 (2008)

87. P. Olver, *Applications of Lie Groups to Differntial Equations* (Springer, Berlin, 1986)

88. K. Painter, J. Sherratt, Modelling the movement of interacting cell populations. J. Theor. Biol. **225**, 327–339 (2003)

89. E. Pennestrì, V. Rossi, P. Salvini, P. Valentini, Review and comparison of dry friction force models. Nonlinear Dyn. **83**(4), 1785–1801 (2016)

90. V. Petrov, S. Scott, K. Showalter, Mixed-mode oscillations in chemical systems. J. Chem. Phys. **97**, 6191–6198 (1992)

91. L. Prandtl, Über Flüssigkeitsbewegung bei sehr kleiner Reibung, in *Verhandlungen des dritten internationalen Mathematiker-Kongresses in Heidelberg 1904* (ed.) by A. Krazer (Teubner, Leipzig, 1905), pp. 484–491

92. E. Rabinowicz, Stick and slip. Sci. Am. **194**(5), 109–119 (1956)

93. J. Rankin, M. Desroches, B. Krauskopf, M. Lowenberg, Canard cycles in aircraft ground dynamics. Nonlinear Dyn. **66**, 681–688 (2011)

94. K.L. Roberts, J. Rubin, M. Wechselberger, Averaging, folded singularities and torus canards: explaining transitions between bursting and spiking in a coupled neuron model. SIAM J. Appl. Dyn. Syst. **14**, 1808–1844 (2015)

95. M. Roussel, S. Fraser, Invariant manifold methods for metabolic model reduction. Chaos **11**, 196–206 (2001)

96. B. Sandstede, Stability of travelling waves, in *Handbook of Dynamical Systems II* (ed.) by B. Fiedler, (Elsevier, Amsterdam, 2002)

97. S. Schecter, P. Szmolyan, Composite waves in Dafermos regularisation. J. Dyn. Diff. Equat. **16**, 847–867 (2004)

98. S. Scott, *Chemical Chaos* (Oxford Science, Oxford, 1991)

99. L. Segel, M. Slemrod, The quasi-steady-state assumption: a case study in perturbation. SIAM Rev. **31**, 446–477 (1989)

100. M.A. Shishkova, Study of a system of differential equations with a small parameter at the highest derivatives. Dokl. Akad. Nauk. SSSR **209**, 576–579 (1973)

101. N. Steenrod, *The Topology of Fibre Bundles* (Princeton University, Princeton, 1951)

102. P. Szmolyan, M. Wechselberger, Canards in $\mathbb{R}^3$. J. Differ. Equ. **177**(2), 419–453 (2001)

103. P. Szmolyan, M. Wechselberger, Relaxation oscillations in $\mathbb{R}^3$. J. Differ. Equ. **200**(1), 69–104 (2004)

104. A.N. Tikhonov, Systems of differential equations containing small parameters in the derivatives. Matematicheskii Sbornik **73**(3), 575–586 (1952)

105. B. van der Pol, A theory of the amplitude of free and forced triode vibrations. Radio Rev. **11**, 701–710 (1920)

106. B. van der Pol, On 'relaxation-oscillations'. Lond. Edinb. Dublin Philos. Mag. J. Sci. Ser. 7 **2**(11), 978–992 (1926)

107. A.B. Vasileva, On the development of singular perturbation theory at Moscow State University and elsewhere. SIAM Rev. **36**, 440–452 (1994)

108. T. Vo, M. Wechselberger, Canards of folded saddle-node type I. SIAM J. Math. Anal. **47**, 3235–3283 (2015)

109. T. Vo, R. Bertram, M. Wechselberger, Multiple geometric viewpoints of mixed mode dynamics associated with pseudoplateau bursting. SIAM J. Appl. Dyn. Syst. **12**(2), 789–830 (2013)

110. M. Wechselberger, Existence and bifurcation of canards in $\mathbb{R}^3$ in the case of a folded node. SIAM J. Appl. Dyn. Syst. **4**, 101–139 (2005)

111. M. Wechselberger, À propos de canards (Apropos canards). Trans. Am. Math. Soc. **364**(6), 3289–3309 (2012)

112. M. Wechselberger, G. Pettet, Folds, canards and shocks in advection-reaction-diffusion models. Nonlinearity **23**, 1949–1969 (2010)

113. M. Wechselberger, J. Mitry, J. Rinzel, Canard theory and excitability, in *Nonautonomous Dynamical Systems in the Life Sciences* (Springer, Berlin, 2013), pp. 89–132

114. W. Whiteman, A. Ferri, Displacement-dependent dry friction damping of a beam-like structure. J. Sound Vib. **198**, 313–329 (1996)

115. S. Wieczorek, P. Ashwin, C. Luke, P. Cox, Excitability and ramped systems: the compost-bomb instability. Proc. R. Soc. Lond. A Math. Phys. Eng. Sci. **467**, 1243–1269 (2011)

116. A. Zagaris, H. Kaper, T. Kaper, Two perspectives on reduction of ordinary differential equations. Math. Nachr. **278**, 1629–1642 (2005)

117. A. Zagaris, C. Vanderkerckhove, W. Gear, T. Kaper, I. Kevrekidis, Stability and stabilization of the constrained run schemes for equation-free projection to a slow manifold. DCDS-A **32**(8), 2759–2803 (2012)